to stand and stare

To Bill

who preferred to do his staring
while sitting on my foot

ANDREW TIMOTHY O'BRIEN

to stand and stare

How to garden while doing next to nothing

Contents

Flowers
(summer)

Fruits
(autumn)

A garden of intention

Let's get one thing straight. Your garden will get along quite well without you. It just won't be your garden. It will be that piece of land where once a garden was, and there is no tragedy in that. Every square inch of our planet's surface is trying to get back to a state where it feels at ease with itself, a way of being from which forestry and agriculture, road building, town planning and, yes, even gardening continually strive to hold it back. This is not whimsy. This is science.

I can see it happening in my lawn, in my flowerbeds and borders and, particularly, in the more forgotten corners of the garden where plants I don't remember inviting in make themselves at home, sinking roots deep down into soil while reaching ever skyward to gather in the sun's largesse. I plant roses and paeonies, giant scabious and salvias, hardy geraniums and bold, bright geums – I water and feed them, stake them against wind and rain, deadhead and chop them to prolong the floral show. And as I labour, unseen by me, someone else plants nettle and cleaver, herb robert and wood avens, wild strawberry and creeping buttercup, foxglove, dandelion, forget-me-not, and dock – each will grow, flower, and set seed, and all of this while asking nothing of the human gardener. Ash tree seedlings will appear suddenly as if from nowhere, now six inches, now six feet (never turn your back on an ash seedling, it will be a rangy sapling tree before you know it), blackberries reach out to snag me across paths, defying summer drought to grow surely a foot a day with no need for the watering can without which my sweet peas would transform as I watch into a crackling tapestry, warp of dried raffia, weft of crinkled biscuit.

To please pollinating insects, I mothball the mower for several weeks and give the lawn leave to run to seed. Hoverflies and bees skim merrily from daisies to clover, selfheal and ribwort plantain. Blackthorn seizes the opportunity to rouse questing roots sent outward on previous excursions from the boundary hedge, and diminutive sloe bushes shoot upward through the turf. Bramble primocanes, having reached the peak of their trajectory, arch and plummet, taking root where they touch the ground. My garden wants to be a woodland, and seemingly the only thing in its way is me.

A garden is created out of intention and held in that state by force of will. It involves partnership and compromise, a coming together of human and nature. It can be a beautiful relationship, or one fraught with frustration on both sides.

En route to woodland lies scrub – that place of scrappy, shrubby growth, not quite meadow nor yet quite forest, where tussocky grasses and wildflowers exist in a jittery kind of half truce with thistle and gorse, bramble, and briar. This is a temporal, as well as a physical reality for the land out of which my garden is carved, within the biome of the temperate deciduous forest, and one of its cardinal agents of change seems to be compost. Long before you or I were exhorted to reduce, reuse, and recycle, nature was composting, and doing it with an efficiency that not only disdains our own half-hearted attempts but lies at the heart of a process that sees bare, inhospitable rock become host to rich communities of life; a transformation begun by the first doughty, pioneering plant, root winkled into a rocky crag, dropping leaves and succumbing itself to the passage of time; environmental enzymes and microscopic organisms breaking down vegetative tissue, the parts combining with minerals weathered from base rock to form a scanty soil just sufficiently rich to encourage tentative communities of slightly greater diversity. And so it builds, from cold stone to grassland, from grassland to scrub, from scrubland to forest.

Every gardener has a little misplaced bravado inside; palm outstretched against the approaching green tide like England's King Canute, constantly anticipating the overwhelming waves that we know deep within ourselves we're powerless to resist. We hold the land at a tipping point, sensing its deep desire to transform itself but unwilling or unable to embrace the wilderness, to welcome the wildness into our domestic space. The dynamic tension this creates lies at the heart of our conflicted relationship with nature and the manner in which she manifests her will in our gardens, a phenomenon intruding upon our notice through the presence of those plants we've come to think of as weeds. An entire industry has sprung up in order to educate us in regard to which plants are desirable and which we should dedicate our spare time to eradicating from the garden, towards the furtherance of which noble effort they just happen to be in the position to sell us the requisite weaponry. It's a phony war, and it's exhausting. What if we were to ignore the accepted wisdoms and surrender to nature, just a little? Give in to the overwhelm and roll joyfully where the verdant tide takes us.

A garden is never, as frequently and lazily opined, all about the plants. Neither is it centred on the people who will spend time within its bounds. Not entirely natural, nor wholly a construction of human artifice, rather the garden is a collaborative effort, a coming together with anything from a bashful kiss to a collision. Fundamental to the way in which we garden will be quite how we view our own selves in relation to the natural world – do we feel part of nature, or is it something we need to harness and tame? Do we revel in the welcoming embrace of the elemental, or does the notion of wilderness fill us with fear and trepidation? All of which can sound a bit high-falutin, but it comes down to this. Next time you find yourself weeding, ask yourself, why? For what reason do you go to such lengths in pulling out plants that grow so well on their own, while the expensive garden centre specimens greedily demand your time and attention before they consent to perform, or hold over you the imminent threat of their demise? We find ourselves unquestioningly perpetuating a narrative that is partly to do with control, partly about imposing our will upon the landscape, partly even about those notions of dominion that descend from the book of Genesis, all the while overlooking – in the excitement of being given authority over natural resource – the obligation to ensure that the earth is "replenished"[1]. To be good stewards, rather than careless consumers.

And so, where nature is to be tamed, we clip, and mow, we bend and tie, train, and prune. This tends to lead to very formal gardening styles, tightly-clipped hedges, and topiary – that most telling of horticultural arts, where the gardener imposes their will upon the very form a plant can take for no other reason than the delight afforded to the onlooker at such incontestable proof of man's mastery over nature. There's something undeniably pleasing about such discipline and artistry, but while enjoying the tight, green curves of a topiarized peacock, it's as well to remember how much work is required to hold a hedge in this form, and how quickly the image will flicker and dissolve should the gardener's attention be distracted. Nature, it seems, will suffer herself to be subdued or even commodified, but only for a season.

But pause with me here a moment as I sit in a shady spot of my sun-parched, summer-ravaged, messy garden, reflecting on the changing nature of the partnership in question. Stop. Breathe (*in… hold… out… waves lapping shore*); swoosh of passing traffic on the London road, a plane overhead bound for Gatwick, neighbours' children playing, the cooing of collared doves and the chirping of sparrows. The slight cooling breeze of this morning has given way to soaring temperatures, and the garden crisps almost audibly, flowers fading faster than I'd like, bright purple artichokes turning dull brown. There is plenty here that needs taming and tying in just now, but not today. It can wait. I can let it go – *let it grow* – a while longer and sit in companionable quietude with my garden while thoughts turn over in my mind. I look about, taking in a season's growth, tired stems supporting flowers gone to seed, hedges breaking out of straight lines and intruding joyfully into the spaces between things, and I wonder about the urge to tame a landscape.

It's notable that until we come to the work of influential nineteenth-century horticulturalist and garden designer Gertrude Jekyll, the roll call of renowned landscape architects and designers – William Kent, Lancelot "Capability" Brown, Humphry Repton, even Jekyll's own friend and colleague the Irish gardener and journalist William Robinson – is exclusively male. It's an offence that Catherine Horwood takes pains to rectify in her book *Gardening Women* (2010)[2], at once addressing the discourse of history being overwhelmingly documented by men, and the fact that female gardeners – denied by societal norms any professional standing until the upheaval of the First World War – largely carried out horticultural roles within the domestic sphere. As a man who plays with plants and flowers for a living, I do my utmost to avoid a lazy kind of binary thinking when it comes to the garden. But to acknowledge both masculine and feminine energies at play within this dynamic seems inescapable: the one controlling, defensive, eager to make sweeping change; the other strong, creative, watchful and, for the most part, biding her time. Experience suggests that these energies generally, but not exclusively, align closely with gender identity in their expression – men seem to gravitate towards power tools and feel an irresistible urge to prune things somewhere between knee-height and ground level, while women, regardless of gardening experience, at the very least tend to entertain the opinion that there must be more to horticulture than clear felling everything in sight once or twice a year. As with every creative partnership, the sweet spot from which the most satisfying endeavours flow is found at the point where energies balance.

What does the sweet spot look like? Perhaps something not unlike those ideals laid out in that revolutionary forerunner of our contemporary naturalistic garden style, William Robinson's 1870 book *The Wild Garden*. Perhaps the not wholly dissimilar vibe of the cottage garden, flowers and edibles all jammed in together where space allows, in the style of a thousand humble country plots laid out with necessity rather than art as the guiding principle. It's not beyond the realms of possibility that your sweet spot will take the form of a modern, minimalist space, though it's never wise to underestimate the effort required to hold nature to such acquiescence.

A line of purpose and intention runs between gardener and nature. It's that purpose and intention that transforms the space around your home from a yard into a garden, and it shouldn't matter a fig to anyone but you if your lawn is supremely well manicured or boasts grass so high a diplodocus could move through it undetected. Somewhere along that line, the two parties meet, the point of encounter open to negotiation, each side bringing to the table the resources they can muster – time, money, energy on yours as the gardener, sheer force of will and elemental power on the other. Nature will push you right back on your heels and roll over you if you let her. Speak softly and show no fear and you may just gain her respect.

All relationships require work, but I am convinced that easing into this one would help more people to connect with their gardens at an altogether deeper level. We are creatures that have evolved in concert with a landscape from which we've largely built ourselves out, hiding it behind and beneath layers of brick and concrete and tarmac. Nature waits in our gardens for us to exit through the back door and begin a conversation, but through fear or ignorance we rarely make the time to tell the space what we require of it – and listen for what it might want back in return. Instead, we look to others for easy solutions to an *al fresco* decorating quandary, perhaps alighting on a preferred look and feel or, more likely, stumbling by default into our own gardening groove, informed by snippets of information incidentally acquired. Often, "style" comes secondary to having a tidy space where the plants we buy do what they're supposed to, and don't embarrass us by ostentatiously expiring just before the in-laws come over for a weekend barbecue.

We can pilfer ideas from gardens visited, the pages of glossy magazines or what we see of the Chelsea Flower Show on the television, but the vast majority of our gardens evolve organically, being subjected to the occasional minor cosmetic tweak rather than a root and branch reworking of the entire space – a process that considers how we live, move, value, and relate to the world around us as it intersects with our home. They transform gradually from the spaces they once were, often laid out for the convenience and taste of a previous occupier, into a garden over which with luck we can claim some degree of meaningful ownership beyond those rights conferred by title deed – extending a flowerbed here, erecting a shed in there, clearing a space for the trampoline or the rotary clothesline.

How many of us approach this area of our homes with that purpose and intention already mentioned when so often our time-poor daily experience leads us to view gardening as a series of chores to be ticked off? There's the mowing, the weeding, the autumn leaf blowing, the desperate squeezing of hastily bought bedding plants into window boxes and hanging baskets each spring so that the neighbours can spend the next few months watching them slowly die as daily, unheedingly, we pass them by in one long drawn-out motion blur. And so, we go with the line of least resistance and adopt the consensus, our gardens becoming shaped less by questions of aesthetics or ideals or thought for how we might want to live, than by market forces. We impose order because we're told that's how to garden. We spray and we feed and we irrigate because we're told what to buy and what to kill, and what to make unnaturally green. We become pawns in this area as in so many others, with insult added to injury by the sheer effort it takes to maintain our polite but boring gardens. Well, thanks, but no thanks. You'll pardon me if I don't want to play. If I choose not to drive to the garden centre, hand over my bank card for a trolley full of chemicals, then spend my weekend trundling up and down the lawn or wandering in and out of flowerbeds dispensing and spraying in an attempt to maintain a standard which, if I stand still for long enough, I realize isn't even my own.

A paradigm shift is called for, the rejection of a stance that sees the natural world as a threat in favour of one where humankind and nature are partners – and even here, we should be aware of a kind of self-deception that places both parties on an equal footing, when any reasoned consideration quickly establishes us as the rather noisy and troublesome child. But since gardens are illusions, places in which we are willingly invited to suspend disbelief for a while, we should at least allow ourselves the delusion for the time it takes us to wander between the flowerbeds.

As we search for this elusive sweet spot, we should be wary of behaving as if gardening, by definition, is all about activity. Certainly, there's always plenty that can be done, but action in the absence of a plan can lead to a rather fruitless (if tidy) space, and a plan without purpose ultimately fails to deliver satisfaction. Time is never wasted when taken to discover the *why* that lies behind any garden you want to make, even if that means standing about lost in thought as bindweed winds its way around your legs and your neighbours complain in loud stage whispers about the tendrils reaching over the fence and snagging the umbrellas in their Campari.

If you really want to garden while doing next to nothing (one assumes you have at least a passing interest, since you've picked up the book), let the garden revert to woodland. Watch how its character shifts as nature gradually reclaims the space. Brave tangled growth to tramp the paths foxes will make through your wilderness – foxes are good at making paths, as are the badgers that might move in as you allow your space to rewild. Spend as much time with lowly mosses and lichens as you do with the loftier heights of holly, ash, hazel, and birch. Begin to number woodlice and worms among your friends. And so, the promise of this book's title is discharged before we're even past the introduction.

Going full tree-hugger could be a little extreme for some, but the ability to picture what that parcel of land could become – *wants to* become – if left to its own devices is fundamental to a fulfilling relationship with your garden, one that brings joy and escape and a sense of peace, rather than frustration, guilt, and a nagging feeling of overwhelm. If becoming a gardener is about entering into a partnership, then it makes sense to get to know who you're climbing into bed with, and while the garden doesn't demand your activity, it does require your *presence*. How you choose to show up in that space, and what you choose to bring into being – whether a flower-filled wildlife haven, a sleek tribute to modernism, or something in between – is entirely up to you. But given finite resources of time, of money, of energy; given also the pressures on our environment at both local and global levels, perhaps a more low-intervention style of gardening – one where we develop the confidence to *let it go and let it grow* – is something that we need to think about embracing more widely.

This might not look exactly like the picture of a well-tended garden you, your fellow gardeners, or those who look upon your space carry around in your heads. But so what? So what if there are weeds (wildflowers – my heart!) in the lawn? So what if you have to push past sprawling perennials as they tumble haphazardly out of the flowerbeds and flop over paths? As well as extending our empathy outward towards nature, we need unashamedly to reconnect with our own preferences, needs, and desires; reclaim the inherent joy of surrounding ourselves with abundance; and stop gardening to please other people. The emotional dissonance so many of us experience when it comes to our garden – that it won't behave as it should, that we just don't have enough time to tend to it, that we don't even know where to start – has at its root a mistaken belief that we are somehow apart from, even above, nature herself. If we can restore this relationship then all our busyness in the garden begins to look less like work and more like time spent with a good, if slightly mercurial friend, with positive benefits of health and wellbeing for both parties.

When it came to writing this book, I didn't think the world needed another "How To Garden" title – there's a wealth of information out there expounding upon the many tasks that it's all too easy to make your garden about. But there's not so much about how you might like to *be* when you're out there, at one with the plants and the wildlife and the weather. I've come to appreciate that an understanding of natural processes is the key to accessing the transformative power of the garden, and replacing feelings of confusion, overwhelm, and stress with focus, a sense of inner peace, and an increased facility to deal with what life throws at us on a daily basis.

With a view to this, over the next few hundred pages, I'm going to invite you to think like a plant. And we're going to start from the ground up.

ROOTS

to nature

It always begins this way. Unseen and in the dark; an emerging, a finding of bearings, a taking of stock. A pause – preliminary to an outpouring of energy, a persistent and questing spirit let loose upon a search for new ground, fresh possibilities, untapped potential. An opportunity to get it right from the start, to create firm and sure foundations, to discover allies along the way and identify points of contact both for now and for the days to come. A vocational calling to be the provider of nourishment and support, without reward, recompense, or fanfare. Invisible. Essential.

We do this, you and I. We start at the bottom and work our way up toward the sun. The garden, beneath the soil. The gardener, in winter.

Wonderland

Winter, and the garden has gone to sleep. Ask anyone.

Hauling myself out of bed I drag my bones groggily downstairs to fire up the coffee machine, before sticking my head out of the back door. Misty. Damp. Chilly, rather than bitingly cold. The first frost of the year has visited, befogging windscreens and prettifying foliage. It's not a heavy frost, but it'll do for now, and I dearly hope it's a sign of things to come. We need a good, hard winter – one that calls for scarves and bobble-hats rather than umbrellas and galoshes.

I want it cold and clear and crisp. I want rosy cheeks and tingling toes, ice in my beard and a flask in my pocket. I want snow, and winter sun, and walks through silent, white blanketed fields to pubs with open fires, mulled spice wine, hearty food, friends and laughter. I want sledging and snowball fights on the way home, slippers and a good book, cosy untaxing movies on the television and the gentle patter of large snowflakes falling softly outside. And when I wake in the morning, I want hoar frosts and ground frosts and the garden transformed into a storybook tableau, like the Christmas department store windows I saw as a child. And if a reindeer should find its way there, so much the better. But I'll settle for cold, and clear, and crisp.

"I suppose you're at a bit of a loose end, now?" the cheery but infuriating comment at which those who garden for a living must learn to bite their tongues all winter long, the better to preserve the fabric of society. To offer up in response a pained smile, or an impromptu lecture? The garden we're accustomed to admire doesn't just happen, magically, in spring. The garden we like to see – the tamed, trained, trussed up and tractable creation, the garden given licence to flaunt its floral wares just so, to hint delicately at the potential of its fecundity but never to frighten with the full force of its desire to grow and to keep on growing, to multiply, to clamber and climb over anything in its path – that garden requires the presence of some poor sod muffled up to the eyeballs throughout the colder months, pushing barrows of manure and chipped bark about the place, weeding and mulching, pruning and tying in, planting bulbs, constructing plant supports and mending fences, turning compost, and lighting bonfires as much to keep warm as to perform obeisance to any horticultural imperative. Never mind that such a person would almost certainly rather be out here all winter long than stuck indoors piloting a desk by the eerie blue glow of a computer screen – for all the skill of plant breeders down the ages, the self-pruning rose or apple tree has yet to be developed. It's a happy coincidence that the garden we like to see requires work to be done over winter, and that there are those who find joy and purpose in such occupation.

Nine times out of ten, I go with the pained smile.

If we travelled around the sun on a planet spinning about an imaginary line perpendicular to its orbit, the seasons would be less differentiated than they are. But the Earth's axis lies at a jaunty 23.4 degrees – our home being ridden through space around a blazing star while pulling an impressive and unceasing wheelie. And so, on any given day, one hemisphere is tilted markedly further away from the warmth of the sun than the other, allowing winter to happen in the temperate north while the south enjoys summer, quite the reverse six months later, and something resembling spring and autumn during the intermediary stages. The climate crisis might be monkeying about with temperature and weather events, but nothing short of a collision with the kind of massive asteroid that's likely to have knocked the planet slightly out of kilter in the first place (she's called Theia, is rumoured to have existed about 4.5 billion years ago and likely persists in some form deep within our planet's crust) is going to change the effect that this seasonal change has on the length of daylight from one month to the next. And so, at the appointed time and with their distinct moods and characteristics, one season succeeds the next – most notably in polar and temperate regions, though hardly at all as you approach the equator – and here we are in winter; dark, chilly, in need of hot chocolate and wondering if we should stay out here and tackle the garden or go back inside and shut the door on it until spring.

Fond as we are of our lawns and flowerbeds, we British are at heart a nation of fair-weather gardeners. It's hard not to conclude that we're only interested in what our gardens have to offer us when the sun shines, when we can kick back with a glass of something refreshing, watching fluffy clouds scud across clear blue skies while the kids and the dog dash about in the fresh air and maybe, if we're feeling active, indulge in a spot of light deadheading or – heaven preserve us – a little weeding. But at the first sniff of inclement weather, we head indoors. Once the October school holidays are behind us, you can forget about going into the garden unless it's to get something from the shed or the garage. Then we'll dash as quickly as we can from the house to our destination and back, with barely a glance at what might be going on outside. We act as though winter catches us off-guard every time, for which there's really no excuse; for all we love to complain about the skittishness of the weather, winter still follows autumn as autumn succeeds summer and summer, in its turn, spring, and all this despite the signs of seasonal creep identified as yet another casualty of the climate crisis. This being the case we should be able both to predict winter's arrival and to prepare ourselves for enjoying our gardens through the cooler, shorter days at the darker end of the year – and yet it seems beyond us. But until we learn to embrace this time of year, to picture ourselves as much a part of the winter garden with its soft ground, stark silhouettes, and jewel-bright berries as we do the summer garden of hazy afternoons among bold daisies and sparkling grasses, we're going to deny ourselves access to this seasonal wonderland. Indeed, it's a symbol of privilege to be able to hold our treasures so lightly; to scorn our soggy winter gardens while we huddle indoors by the fire and the telly, when by rights we should be out there every day of the year, treading each square metre of our lot with gratitude. And since nobody denies the value of land – whether we seek to own it, exploit it, tend it with respect and understanding, or merely enjoy its many beneficial contributions to our mental, physical, and spiritual health – why we should choose to absent ourselves from any stretch of accessible ground for at least half the time we spend on the planet seems, at the very least, hard to fathom.

We can just about bring ourselves to cope with autumn by enlisting the aid of shiny conkers, welly boots, and pumpkin spice lattes, but Hallowe'en and Bonfire Night mark a very real end to our experience of the outdoors for the year – at least in a manner which involves either purpose or intent. As everything in the natural world appears to slow down and slumber, perhaps we could be forgiven for wondering why we should spend time outside. What – if anything – is going on in the garden in winter, and why would it require the presence of a person?

Winter moments spent in quiet meditation of our beds, borders, and window boxes often reveal that the garden's own level of activity has not entirely stalled, but rather is quietly ticking over. Being the season when that part of the globe over which it exerts influence is turned away from the sun, winter is by definition both dark and cold – a combination with particular consequence for the plants in the garden, all of which need light and warmth in order to produce the necessary energy for growth. And while there might be an abundance of the water that's also required before the transformational magic of photosynthesis can take place (see *A free lunch*, page 123), assuming snow and ice haven't rendered it temporarily unavailable, the growth rate slows materially, if not quite to a standstill across the garden as a whole, then very nearly so.

And after all the frenetic activity of summer ripening and abundance, all the showy drama of autumnal transformations, the pause is as welcome as it is needed, as much for the gardener as for the plants they tend. A slowing of pace that brings the chance to take stock, plants held in check, all above-ground developments frozen until the spring. Except for a few staunch evergreens keeping the show going, everything would appear to have entered its period of hibernal rest. It's worth remembering however that with plants, what we can see is only half the story. Beneath our feet, energy is being held deep within root systems, insulated from the cold, waiting for more propitious conditions for the putting forth of new shoots and leaves, the ground acting as a bank vault into which the garden has buried away its treasure against the better days. The soil itself is so much more than a body of inert dust and mud into which we sink our plants and the foundations for our houses, and it would be a mistake to assume all subterranean activity stops over winter. While drowsiness reigns up top, microorganisms continue to grow in the root zone – even when the ground is frozen – albeit at a slower rate than in the warmer months. There's even some evidence[3] that certain members of the soil microbial community are at their busiest in the depth of winter under a cover of snow, only slackening off, job done for the year, during the spring thaw. Viewed from the perspective of the garden ecosystem, the role of these particular bacteria is to recycle nitrogen – a key component both of plant proteins and the substance known as chlorophyll responsible for driving the process of photosynthesis – and it should come as no surprise that they would spend the winter laying down as much of that element as possible, ready for roots to absorb when spring's brighter days and higher temperatures act as both call and catalyst to growth.

But while we're shivering into our coats and wondering when we can go back indoors, the plants in our gardens – quite literally rooted to the spot – have evolved their own strategies to cope with the cold. Annuals grow, flower, and set seed all within the space of a single season and, while most will die before winter even gets a grip upon the garden, their legacy is coded into the genetic information stored in seeds – some of which require a period of chilling before they will allow themselves to germinate – scattered across the soil or transported to new locations by birds and animals. Biennials such as foxgloves and teasels keep a low profile during their first year, hugging the ground in mounds of sturdy foliage throughout the winter, only producing flowers and ultimately seeds in their second growing season after which, like annuals, they expire. Those plants with real staying power – the ones that persist from year to year – are the perennials, which might be either evergreen (see *Ever green*, page 100) or deciduous, and either could be shrubby or herbaceous; the shrubs maintaining an enduring framework of strong woody stems, while the soft green stems of their herbaceous neighbours must either be strong enough to withstand low temperatures, or die back to the ground each winter to be replaced by new growth in the spring.

Deciduous plants have learned to cope with the winter by temporarily dispossessing themselves of the organs that would otherwise be vulnerable to the cold – most notably their leaves, though stems too in the case of herbaceous plants, which are sealed off at the base and left to stand or fall as wind and weather may take them. This release contributes much to the gardenscape as the year draws to a close, providing hibernacula within which insects may drowse away the winter in relative security, as well as food for the creatures and fungi that depend for survival on decomposing plant material. Meanwhile, the plants – protected by an insulating layer of woody tissue or blanket of soil – are spared from the worst ravishes of frost and freezing; the ill effects of the water so critical for survival as it turns to ice, becoming the enemy within.

Most substances contract as they change state from a liquid to a solid, but water only follows this general rule to a point. Below 4°C (39.2°F), water begins to expand again, further by about 9 per cent when frozen. This is good news for fish in winter as it means the ice on lakes and ponds, being less dense than water, floats to the surface allowing aquatic life below to persist through the colder months. But the implications for plants are less positive as cells rupture and dehydrate, ironically through the unavailability of the same water locked into the keen-edged ice crystals now wreaking havoc upon the cellular membranes within the tissue. Who could blame the oak or the ash for shuffling off its leaves in autumn, or the delphinium for shutting off all above-ground operations, the better to avoid such injury? And yet the hellebore appears to flaunt a kind of foolhardy bravado in the face of the coldest weather, holding on to both leaves and soft, green stems, bent right to the ground in the frost and then, miraculously, springing back upright as the temperature rises through the day, its own winter adaptations relying heavily on antifreeze proteins held within its cells (that other favourite of the winter garden, the snowdrop, uses a similar mechanism).

While the hellebore is a perennial both evergreen and herbaceous – a combination demanding our respect – the snowdrop is a bulbous perennial, which is quite the horse of a different colour. Certain plants bide their time in underground storage organs, emerging only upon the arrival of conditions most favourable to their flourishing and multiplication. In the case of bulbous plants not only the roots, but also modified forms of the stem – now a flattened disc referred to by botanists as the basal plate – and the leaves, or the lower portions of leaves, persist beneath the ground, the latter forming fleshy "scales" surrounding the nascent shoot that will eventually force its way up through the soil, and containing the energy reserves that will fuel that journey into the light. The scales of some bulbs – tulips, alliums, narcissus, all waiting winterlong a few inches beneath the surface – are surrounded by a protective papery sheath; others – lilies, or fritillaries, for example – have naked scales. The strategy is identical in either case; to wait out adverse conditions inside a self-contained capsule, a tiny parcel of potential energy, primed to go off with a floral explosion at the appointed time.

It's a phenomenon increasingly lacking in the power to surprise – the ability of a section of human society to turn to its own economic benefit, over the course of a century or so, a solution that will have taken nature millennia to evolve for the benefit of plants, but it could well be argued that the bulb's properties of stability and resilience helped fuel the Tulip Mania of the Dutch Republic; a seventeenth-century speculative dash for botanical bitcoin. A self-contained and portable asset that would store well during long voyages in the holds of the ships of the Dutch East and West India Companies was suited to trade throughout the colonial empire, though by the time the mania hit its height money would change hands several times without a single bulb being seen by speculators. Such frenzied over-stimulation of the market went the way of all bubbles and burst, but the trade in bulbs remains strong to this day; billions of them grown, lifted, and packed to be sold and planted around the world each year. Where would our spring gardens be without tulips and daffodils, our summer borders without lilies? For that matter, where would our winters be without the plant catalogue's promise of the brightness and colour to come?

But not everything sold as a bulb is a bulb. Any plant that folds itself away into an underground capsule with the purpose of waiting for more auspicious conditions falls under the heading geophyte, but the retail nursery trade, being in the business of selling things, seems wisely to have decided that this kind of technical term lacks a certain catchiness. And so, while browsing the pages for flowers with which to populate our borders in the new year, we're liable to encounter a variety of distinct and different perennating storage organs (a phrase even less widely used), all lumped under the one, albeit slightly misleading, heading. There are those that, like true bulbs, employ some modification of the plant's stem: staunch and solid corms, such as crocus and cyclamen, sharing with bulbs a flat basal plate but lacking the concentric layers of fleshy leaves; long and lumpy stem tubers with their numerous growing points or "buds" familiarly encountered as the eyes on your potato; starchy, horizontally-growing rhizomes, represented in the kitchen by roots of ginger and turmeric, and in the flower garden also by ginger, as well as the bearded iris. There are, too, the swollen root tubers of dahlia, daylily, and sweet potato, each demonstrating the variety of invention within the plant kingdom when it comes to stockpiling energy over the leaner months.

All this slumbering, unseen in the ground beneath our feet, waiting for the right combination of conditions before rousing themselves back into growth; a quietude that offers us opportunity. How we spend that opportunity over winter is a thing that defines our relationship with the garden over the year to come: we could award ourselves time off, closing the door till spring and allowing it to get on with things by itself; or we could follow so much of the prevailing garden advice trotted out online and in the Sunday papers, by "putting the garden to bed for the winter". The first option might seem like the clear favourite when looking for a low-maintenance approach, but what's really on offer here is postponement followed by panic, and not the kind of light-touch gardening we might at first have imagined. Because as soon as spring comes around, and always assuming we're not going for the full wilderness look, much of last year's growth will need to be cut away before new shoots emerge among the old, to say nothing of the general husbandry of both plants and soil, and doing this across a whole garden while everything around is bursting into life becomes a race the gardener is destined to lose. The second option sounds much tidier – cutting plants back, raking leaves away and generally making everything look neat and respectable – but it's an undeniably high-maintenance approach, and not a little clinical.

More than this, what both strategies have in common is the total absence of the gardener from the garden throughout an entire season, and it's hard to invest any worthwhile undertaking with the intention it deserves when attempting to do so by remote. As with all the best relationships, our presence, if not absolutely required, is certainly desirable.

What if there were a third way? There is.

*

But step into the winter garden with me and stand here a while. Sitting is good; sitting still even better, and something that can be encouraged by the curiously often-overlooked expedient of punctuating the journey around and through the garden with things on which to sit. When we sit well, rather than slumping – with backs straight and shoulders dropped – we breathe more deeply, we slow down, relax, and invite a state of calm contemplation. We might even be tempted to drift off into that slumber through which thoughts are free to form connections unhindered by the kind of narrow policing our conscious mind seeks to impose upon our mental processes. There will be time for much of this over winter, but now, as a signal – to ourselves, as much as anyone else – that we mean to get serious about this gardening business, we can adopt a pose that models that sense of attentiveness and potential, and stand. There's nowhere to sit out here just now without getting a soggy arse anyway.

*

Stepping outside is the first ground to be claimed in that battle we have with our inner resistance to gardening during the winter months. There's something liberating – invigorating, even – about leaving behind the central heating and the dry air, closing the door upon the chatter of the television or the murmuring of the radio, and heading out into the world beyond the back door. Plunging directly into snowstorm or downpour, any al fresco meditations may be short lived, but depending upon our speed of exit from the house, our sartorial preparedness, and the disposition of the weather, we could be some way into the garden before becoming fully aware of any difference in the immediate surroundings. Assuming relatively benign skies, a chill in the air, and a softness to the earth in the company of a thick jumper and a hat hastily crammed onto the head, it's possible to envisage standing here, just within the bounds of the garden, for long enough to enter into that receptiveness where the small details of our environment – the kind of thing that we miss on the daily dash from doorway to car to garage and back again – are permitted to intrude upon our notice. Make a habit of this, and we begin to cultivate that unbroken line, that continuity of purpose and intention that we can bring to the garden through each season of the year, a continuum threatened by our own desire to hibernate.

Because almost all that is required of us as gardeners is that we *pay attention*. Now we notice the cold, of course, but also the damp – the air around seems fat with water, hanging heavily about us and running down stems and along twigs, clinging prettily from the tips of leaves before dropping to the ground, and we're sucking in quite different stuff to that which we breathe indoors at this time of year – feeling every inbreath as a hydration, every outbreath a child's game; we are a dragon, a chimney, a cloud machine, our exhalations visible as vapour, merging with the moisture of this winter's day. A billowing of breath and the dampness of droplets pooling upon earth that smells so healthful, so fresh and uplifting you could almost take a bite and imagine if you did it would taste like – what? Something good, something wholesome, the best kind of chocolate cake that's not too sweet and still has a bitter tang of cocoa; a really good brownie. We're not going to eat the soil, we're not even going to dig it if we can avoid it and, where possible, we'll try not to walk over it too much. The kindest thing to do over the coldest months would be to draw a thick covering of mulch over it, but first we need to get to it, and there is much of last year's garden in the way; a blanket of broken stems and fallen leaves as though nature has taken matters into her own hands and seen fit to create from found objects an insulating layer to warm and nourish the earth. Which is exactly what has happened, because the garden, never afraid to seize the initiative, has no need to wait upon the gardener, whose arrival is now announced as the excited chirrup of a companionable robin is joined by the rhythmic squeak of a poorly lubricated barrow wheel, the two winding upward in counterpoint into the silent vastness of a bright, white winter sky.

As a jobbing gardener there are things that I need to do during the winter months in order at least to give my clients the illusion of a garden under control throughout the rest of the year. But here in my own space, I can, if I want, allow myself a little more freedom. And so the question becomes not so much *how much do I need to do* over winter, but *how much should I be doing*, inviting speculation over whether or not the garden would actually welcome a temporary rest from my well-intentioned ministrations. A period in which to hit the reset button, where birds can gather seeds from flowers left standing way past their prime, and insects can sleep away the colder days in hollow stems that might otherwise long ago have been burnt or composted, and where the soil can breathe a while, untroubled by spade or heavily booted feet. I can narrow down the job list to only those activities for which winter provides a distinct advantage: those that are easier over winter due to improved access around and through the beds, and those that are horticulturally timely, such as pruning and training roses, or planting bare-root plants. But all the while, seeking that third way to garden over winter, a middle path between doing nothing, and doing over much; an approach that acknowledges those aspects of the natural world which are in short supply while the garden slumbers – light, warmth, growth – while taking advantage of the gifts that winter can offer in abundance, chiefly space, time, and clarity. Because if we're serious about finding a way to garden by doing next to nothing, then what better place to start than here – than now, in winter – when so much of what is required can be accomplished in thought as we review the ghost of the garden past and look ahead to the garden yet to come, and the list of that which can't be undertaken without physical effort can be efficiently whittled down to a manageable series of actions. To achieve, in short, a balance of doing, and *being* – the key to which is in the dreaming.

What dreams may come

At the root of how you or I feel about our garden, lie dreams. Beyond how we'd like the space to look, beyond the plants and the path and the patio, our garden dreams are fed by notions of identity and the stories we carry within us about our relationship to the land and, ultimately, to one another. This patch of ground set apart from the outside world by fence or wall or hedge becomes not only a stage on which dreams are played out, but a focus of imaginings in which an ever-present figure is the best version of ourselves, at work, or rest, or play.

On those dark, cold, wet winter's days, when only the most hardy or extraordinarily perverse would venture outside, the job of the gardener is to gather up a towering pile of seed catalogues, good books, and gardening magazines, sit by the fire, and dream next year's garden into being.

There are two strands of kinship that underpin our garden dreaming. First, there's how our gardens relate to the wider landscape, and then, there's how we ourselves relate to the land. And there's a silver thread that dances around and within and between these twin ties, now binding them closely together, now pulling them further apart, that, in all the intricacy of its complex entanglements, begins to form a third thing: a dreamscape of possibilities that starts to emerge as we weave narratives of truth and longing in and around one another, layering story upon story as aspiration becomes expectation, and expectation reaches ever hopefully towards fulfilment which – naturally – remains as elusive as ever; else what would there be left to dream about?

We might not think we know much about how our garden sits within the landscape; we might even consider the question "what is a garden?" so self-evident that it needs no discussion but, really, the key to the one lies within the answer to the other, and I've found it beneficial to my own understanding of how the garden works to invest a little time in reflection hereabouts. It's surely worth at least the time it takes to make, brew, and drink a pot of tea, so put the kettle on, and I'll tell you how I see it.

It's not uncommon to read of human activity described as something "imposed upon the landscape". The Romans accomplished this by laying their famously straight roads across an expanding empire, though aerial photography of the countryside reveals that agriculture began leaving its mark within the soil far earlier, in the pattern of field systems dating back to the Bronze Age. We do it now by carving up the land with our railroads and motorways, laying down business districts and holiday complexes, housing estates, out-of-town shopping centres and all the concrete infrastructure that goes into making our modern world. We have grown accustomed to building the landscape out of our lives, co-opting and capturing it in segments only when its resources can be commodified, and so it can come as a surprise to realize that the wilderness is still there, not merely beyond the city limits, but beneath our streets, springing up from cracks in the kerbside and refusing to relinquish its hold on our flowerbeds and, more quietly but with no less insistence, our minds.

And so, we build houses for ourselves, on parcels of land sized according to our pockets. But even within the most modest of plots, it's not uncommon to leave a little padding within the boundaries; space where we can get at the soil to plant a tree or sow some seeds, to conjure up from the land something buried deep within ourselves and call into being plants in whose presence the house is transformed from a building into a home. Does this sound like a luxury? For all my love of gardening, and the meaning I made from these plant-filled spaces, I have wondered if my passion was little more than a preoccupation with those inessentials afforded only to a privileged few.

Ironically, it was a garden at that most privileged of horticultural events, the Chelsea Flower Show, that introduced me to the work of the Lemon Tree Trust and opened my eyes to the universality of the desire to grow and to garden, irrespective of wealth or circumstance. For the 2018 event, designer Tom Massey transformed a corner of the Royal Chelsea Hospital's Main Avenue into a space highlighting the importance of gardening to refugees and internally displaced persons, many of whom number seeds and plants among the few essential possessions brought with them as they flee towards safety. The show garden itself – testimony to the unexpected beauty often encountered in the camps – was built in consultation with residents of Domiz camp in northern Iraq, where new arrivals are processed as quickly as possible from tents into semi-permanent shelters and given access to breeze blocks, corrugated sheeting, and other building materials with which to fashion for themselves something approximating a home. Space in the camp is at a premium – bedrooms double as dining areas, meals typically eaten seated on the floor, rooms continually repurposed according to need – but many families leave space within their allocated footprint for growing. While those from agrarian backgrounds might naturally be inclined to grow edibles with which to supplement their diet, this hardly explains the roses and pelargoniums, marigolds, hollyhocks, daffodils, and the sheer variety of plants that have found a place primarily for their ornamental value. Plants grown to evoke memories of gardens left far behind, kept close to foster a feeling of belonging in a strange land and echoing – in a manner far more visceral than I'd ever had cause to experience – that sense I had of the power of plants and gardens to transform a house from a shelter into a home.

A garden, then, can be thought of as that bit of land we get to call our own for a while, but where our house isn't; where we grow things which have resonance for us, connecting us to our past while evoking notions of family, belonging, and home – themes which fuel our dreams. And still there exists this dualism between what goes on in the garden, and what happens in the landscape within which it's built. Looked at another way, a garden happens when we receive permission to mandate what occurs upon that patch of earth, almost always in close proximity to a house. A title deed or a rental agreement changes hands, and we get to decide how it's going to look, how it will be used, those plants that will grow in its soil and, just as importantly, those that won't receive permission to take root and flourish. Beyond the garden wall nature decrees what grows where but, within those boundaries, we like to think we're in charge.

In reality our gardens do not exist in isolation; although they may feel like little worlds of their own, they maintain a deep tie to the bit of the planet on which they sit, the same landscape that we've temporarily obscured with our built environments. It's a dynamic vibe, frequently overlooked when we envision how we want our gardens to be, but just as overlooked is our own relationship with the land.

When you picture spending time in the garden of your dreams, what is it that you see yourself doing? Wafting gracefully with a trug or a basket and secateurs perhaps, between borders stocked with abundantly flowering rose bushes, or grubbing about happily on hands and knees in the dirt, wresting sweet, earthy parsnips from the soil? Looking around, your gaze could be met by the avenues and fountains of Versailles, the tower and courtyards of Sissinghurst, the pretty, productive jumble of a cottage garden, or the sensible, unpretentious plots of a smallholding – and this being a dream, each can be true at once, roots of heritage, aspiration, and identity running deep through all our garden meditations. Finally, into this heady mix of gardens known or imagined, you stir images imprinted upon your memory while flicking through the pages of glossy lifestyle magazines or scrolling endlessly through social media.

Advertising presents us with images of gardens we're told to want; told we can have. And while the practicalities of wanting and having a particular style of garden is something we'll look at further in *Picture perfect* (see page 160), there's an ingredient at once more abstract and more prone to niggle; the proverbial pea under the mattress. A nagging sensation of discomfiture will trouble the best of relationships, and any disconnect between aspiration and experience is bound to disrupt the harmony we're able to achieve with our garden. It explains why it's often so hard for us to make peace with the space, and is one of the reasons why traditionally we've chosen to subdue our lawns and borders with chemicals, to carve and trim them into compliance and submission. Further, there's a difference between gardening – that I hope to prove is not a difficult thing to do, and one to which the perceived barriers to entry we'd do well to remove – and having the idealized garden, which requires quite another level of commitment, resource, expertise and, yes, privilege. What has introduced this discordant note into our gardening dreams?

We move into a new home with a garden space, and suddenly we have land, though statistically most of us are not "landed" – that is, from the landowning class. Delve far enough into our ancestry and for many the relationship of our forebears to the land will be one marked at the very least by dispossession and disenfranchisement, and quite likely by alienation, forced relocation, and a deep sense of loss.

Of late, rediscovering our connection to the land has become quite the thing to do. Just three or four centuries ago this relationship was self-evident, a truth so universally acknowledged as to be hardly worth the remark. Now, we write books on the importance of rewilding; yes, for the benefit of the planet, for the conservation of plants and fungi and birds and insects and all the critters that live upon the land, all the ecosystems currently under such dire threat, but also for our selves, for the good of our souls. We speak of seeing beyond our built environment, beyond the barriers we have erected to the wilderness that modern-day enlightenment has sought to obfuscate, with which present day post-modern sensibilities seek to re-engage. We cast about ourselves as we hurry to work, emerging from the subway in self-congratulatory mood upon noticing how hairy bittercress and groundsel have, in spite of it all, winkled their roots into some kerbside crack; how ivy, grass, and ribwort plantain have made a home for themselves in some god-awful excuse for soil, amalgam of rotting cigarette butts, drizzle-diluted filth and vulcanized tyre particles, a brave reminder of the world beyond the walls. Plant blind? Not us. We're the new lot, the enlightened ones, the see-ers and the seers, and we preach a gospel of plant-based harmony and inter-dependence with the natural world.

Time was, all this had yet to be lost; the forgetting and the stealing away, an unimaginable future. Privilege confers the option to be oblivious to uncomfortable issues whilst pottering about the garden, but our shared humanity demands of every gardener a pause to stop and think about a relationship to the soil that so many of us take for granted. Because how do you dream of soil and home ground, when the burden of knowledge includes ancestral memories of being evicted from it? Human history is marked by diaspora and disconnection, land grabs of dubious legitimacy enacted in law by the powerful and entire peoples swallowed into economic systems reliant upon enslavement and indenture or rounded up as an inconvenience and forced to assimilate into a settler society. Centuries of violence wrought upon the relationship of human beings to the land, brazenly encoded into law and enshrined within commercial practice, as witnessed by a shameful inventory of policy including, but by no means limited to, the abduction of African people through the transatlantic slave trade, the forced migration of the British working poor from countryside to city following the Inclosure Acts and Highland Clearances, and the genocide of the Indigenous peoples of North America and Canada given false legitimacy by the Indian Removal Act; each action instigated by the European powers of the seventeenth, eighteenth, and nineteenth centuries and perpetuated to this day by an enduring white colonial mindset. Present-day culture wars might seem tame by comparison, but the vitriol poured out on social media in defence of a status quo that denies fault or failing is strongly suggestive of the adage that the more things change, the more they stay the same. And, as we have seen in the work of the Lemon Tree Trust, that failure of international policy and statecraft euphemistically labelled the "refugee crisis" provides an all too real, present-day reminder of both the consequences to and the resilience of people who have been forced to leave the soils of their home.

It would be facile to suggest that the answer to all of humanity's woes lies in the garden, but it's a venue sought by many where the process of healing can begin, arrived at in community or following a solitary pursuit. Ideally, a little of both. And whether the picture of our ideal garden that we hold in our mind's eye is a reminder of a home we've never known, or a vision of one we've yet to create, there will continue to exist for each of us a nagging voice that we can always do a little more, that however unattainable it might seem, maybe there's the hope of getting back home, back to our roots, back to a notion of our selves in all our fullness, through the soil.

The label "dreamer" is rarely applied to anyone as a compliment. Somewhere along the line, it was decided that a preoccupation with dreams is a thing to be denigrated; we speak dismissively of someone having their head stuck in the clouds while showing a marked preference for people who are "down-to-earth". An inventor, by contrast, is a character to be admired – but what's an inventor if not a dreamer who follows through? Where would we be without dreams to fuel invention, discovery, and creativity, or the ability to entertain the improbable in order to stretch the limits of the possible? A gardener needs both to dream and then to make those dreams reality and I'm convinced that, if we're to get the most out of our gardens, we need to spend as much time in our heads as we do on our knees.

And then, there are the *kinds* of dreams. With our peaceful dreams, those soporific reveries where all is sunshine and harmony, our job is to chase down and capture the elements so that, later, they might be incorporated into the waking hours we spend in a space of colour, light, scent, form, and texture.

Work out what brings you joy, peace, serenity, healthfulness – the inclusion of which will help to create a garden where the governing principles might be relaxation and renewal. Just as with people, so with plants; you want to surround yourself with the positive and sparkly ones, those in whose presence you find yourself walking tall, breathing deeply, and regarding the trials of the day with an insouciant detachment, while banishing any that bring you down (this is so much easier to do in your own garden, where nature might allow you at least a 50 per cent say in who you mix with, than with people on the other side of the fence where, most often, you'll have less control).

But our dreams can also be unsettling, and we shouldn't overlook what our subconscious minds might be trying to convey to us through that hazy disquietude. If the garden just described sounded a little too saccharine to provide anything more than a temporary diversion, we can cater to the needs of our capricious and restless natures by embracing, alongside the principles of rest and renewal, the notions of challenge and stimulation. Quite how and how far such creative tension should be included is a matter of taste but its existence within the garden is undeniable; even in the most manicured of green spaces, that element of creative tension exists, or else where would the impetus be to go out and mow the lawn every Sunday?

Even with a looser, more naturalistic look, that tension remains in play between our own vision for the space and how the land wants to be, those plants we want to sow and to tend, and those that the soil wants to grow. This would be niggle enough to be getting along with, but we can add issues of our own identity, history, and circumstance, societal notions of how a garden should look (fashion, the neighbours, your mum), and practical issues arising out of a conflict between how the garden is laid out and how we feel inclined to use and move around within the space. In other words, notions of how we want the garden to feel – or how we want to feel in our garden – and how we want our garden to work, to be, spinning around within our mind in a creative vortex whose internal dynamic allows for one idea to be born out of the mid-maelstrom collision of two others, but which, like a tornado, needs to make landfall in order to have impact on the ground.

Helpfully, the gardening year is divided into seasons and, while it's good to be alive to new possibilities all year round, there are periods that lend themselves more readily to cogitation than to action. The winter months are made for that rare combination of reflection, inspiration gathering, and vision casting that will prove so instructive and fruitful in the months to come. You will find your dreaming self out here in the garden among the fallen leaves and the standing skeletons of summer glory, fresh ideas enthusiastically presenting themselves at the most inopportune moments, finding you half-way up a ladder in the process of wresting a bramble from a tree, or on your hands and knees, face to face with a complex entanglement of things you want cut back hard and things you'd prefer to leave standing. (In the absence of a dedicated scribe, this is where voice-activated digital assistants prove their worth, with many a garden revelation shouted in the vague direction of the phone from inside the undergrowth).

But a more suggestible version of yourself can be encountered indoors, wiling away a wet winter's afternoon by the warmth of the fire, leafing through seed catalogues, gardening magazines, and discovering that the main threat to the advancement of your garden schemes is whether you can record the ideas that present themselves to your mind's eye before you drift away into a doze. It's entirely due to the slippery nature of these visions of a garden-yet-to-come that we need to put structures in place to capture them as, without warning, they appear before us. The wisest of us have ever recognized the power of dreams to influence our wakeful time on the planet, so whatever form your dreamcatcher takes – sketched plans, mood boards, clipping files, scrapbooks, voice memos, wish lists of plants and seeds, or a combination of each of these – it behoves you to keep it to hand if the vision is to have a chance of being fulfilled.

Spring brings with it the first encouraging signs of the garden returning to life, a moment where what is and what might be exist alongside one another in delicate balance. You may emerge from the dark half of the year full of plans to augment the garden's existing glory, to cut it all down and start again from scratch, or something in between. However you intend to go about it, since winter dreams become reality when they hit the ground, an understanding of the soil will provide the firmest foundation.

A handful of dust

Scrape up a handful of soil from the ground and take a close look at what you hold. Such a simple action has the power to make you feel more connected to the earth, or to prompt a desperate urge to wash your hands clean of the besmirching grime. It's perfectly possible to feel both at once. Our relationship to the land is complex, but the land holds the key to our continued existence, and a right relationship with the soil offers a pathway to abundance.

In a most foundational way, my education was profoundly lacking. I skipped biology and the natural sciences at school and concentrated on the arts, my mind preoccupied with loftier concerns than the ground I walked over or the dust I would knock off my boots at the end of the day. Later in life, the evening classes I took to begin reshaping my work around more earthy pursuits taught me to know what kind of soil I had my hands in, how well it would hold on to nutrients, and how it would react to being dug. But it wasn't until proudly flaunting my shiny new horticultural credentials that I was floored by the simplest of questions: what – *really* – is soil? I'm not sure they covered that at college – at least, they never answered a question with quite such an ontological bent. Perhaps I should have listened harder; or asked better questions.

Perhaps that question is best addressed by asking two more: *where* is soil, and *how did it come to be*?

Between the soles of your shoes and the Earth's crust lies a layer of soil – the pedosphere – wrapping the planet around like a blanket: deep, rich, and luxuriant in places; sparse, poor, and arid in others. It is not generic stuff – as any devotee of the forensic crime drama will tell you, a sprinkling of soil carries information as unique as a fingerprint, allowing those in the know to pinpoint its origin to metres rather than miles.

With your handful of dust, a significant proportion of what lies in your palm will comprise base rock, cracked by the climate over countless millennia, gouged by glaciers, and weathered by the elements into tiny particles, each carrying an elemental blend of minerals like an individual signature. The soil at any one point on the planet is inextricably linked to the underlying geology and so, as the layers of bedrock change character from one location to another within the landscape, so too does the soil. In this way, gardening in the south easternmost corner of the United Kingdom where the uplands of the Greensand Ridge intrude upon the wide-open sweep of the Kentish Weald, I may be planting into anything from a soil rich in sandy limestone to a kind of heavy, flint-filled clay, depending on whose garden I'm in on any given day.

But there's more to soil's recipe than rock. To one generous portion of weathered mineral matter, add half a portion of water, the same quantity of air, and top with a sprinkling of organic material (humus) derived from the decaying remains of plant and animal life. Mix well. All life is here, from earthworms to slugs, millipedes to mites, microbes and fungi, the latter in forms variously benign or pathogenic to the plants in our gardens, some existing in a network of reciprocity with the roots of their distant photosynthetic cousins (mycologists now believe that fungi might have more in common with you or I than with plants), others intent upon nefarious burglary, robbing the plants of the contents of their cell walls with no attempt at restitution. There is a city beneath our feet, every bit as complex and interconnected as the urban centres through which we move, in which each citizen has their role to play.

It's this blend of the mineral and the organic that makes soil so fascinating and so vital, its preservation as a resource critical to the continued survival of our species and those with whom we share this place in the cosmos. And should we need reminding how closely our futures are interwoven with the health of our soil, it also has ways of making us happy. Soil-borne bacteria acts upon our brains in much the same way as pharmaceutical antidepressants, increasing levels of serotonin, reducing inflammation and feelings of stress. That wonderfully uplifting smell of damp soil after rain – petrichor – that launched a thousand memes and inspired an episode of *Doctor Who* is more than a product of your own peculiar body chemistry – more than science fiction and rooted in science fact: the smell named for the blood of the gods; the bacteria, *Mycobacterium vaccae*, after a cow from whose dung the first culture was extracted. It's hard to think of anything more down to earth than that.

Add a little water to your handful of dust. Not too much – we're looking for a nice, thick mud – more Glastonbury on a wet year than the stuff that splashes up your back on a soggy bike ride. Now roll it between your palms in a circular motion until you're holding a little brown ball (add a little more water if this helps), then roll this into a nice, fat soil sausage about the width of your hand. What happens to this sausage when you attempt to bend it around into a ring will tell you the kind of soil you have – whether it's full mostly of clay, or sand, or silt. Equipped with this knowledge, you'll be better placed to know what your plants need from you to thrive.

If the sausage bends without breaking, you have a clay soil. If it's reluctant even to hold the sausage shape, let alone be bent around, it contains a high proportion of sand. If it will suffer to be bent a little but begins to disintegrate the further you shape it, it's silt.

People love to complain about clay soil and, though treated with sympathy and care it offers some of the richest and most nutritious growing conditions, the reality is often less than ideal. Clay soil is "heavy" – if you've ever tried to dig one over when it's sodden, or had the bone-jangling experience of trying to thrust a spade into one when it's dry, you'll know why. Clay particles are the smallest of the three soil types and, on the plus side, are particularly good at holding onto nutrients and water molecules. The downside is that this makes the soil prone both to waterlogging and compaction, leading to a lack of air within its structure. Clay soil also has something of a "memory", meaning that working, or even walking on it when wet can have implications in weeks and months to come.

In contrast, the larger mineral particles in a sandy soil allow it to drain far more freely than clay – even in the wettest of winters, plants will never experience the problem of sitting with their roots in water or the accompanying dangers of suffocation and rot. But the speedy downward passage of water through the soil means that nutrients are quickly washed beyond the reach of plant roots, while watering in warm weather can require commitment.

Somewhere between clay and sand in terms of particle size, and sharing something of both their characteristics, lies silt, creating a moderately fertile, well-drained soil – albeit one liable to compaction. But bang in the middle of this trifecta, the unicorn of soil types, is a substance possessing the finest qualities of each: loam.

Traditional wisdom has it that the best way to know what plants will grow well in your garden is to test the soil for its pH, and quite possibly its nutrient balance; certainly a sound and scientific approach. But if such horticultural pedantry seems like a joyless way to garden, you could simply take a good look at what's growing away healthily in your neighbours' gardens and let that inform your planting choice.

Some plants – hardy geraniums and lady's mantle among them – are blithely unconcerned about the kind of soil in which they grow, and gardeners value these reliable, adaptable characters for their ability to contribute to multiple garden situations. Other plants are far more discerning, such as acid-loving rhododendrons that thrive in soils with a pH lower than 7, but turn their toes up over limestone, where the soil tends towards alkaline. In contrast, the sweeping grassland slopes of the South Downs are home to a unique flora especially adapted to their thin chalky soils, including the pyramidal orchid (*Anacamptis pyramidalis*) and other wildflowers that would struggle in more acidic conditions. But whatever their preferred soil type, all plants have in common a short list of basic requirements.

In many ways, plants are not unlike people. They need security, water, air, food, and even friendship, but while you or I might look to a variety of different sources to meet these needs, a plant can get them all from the soil. The yin and yang of good soil exists in the balance between stuff and not-stuff – the particles that give stable anchorage to the roots, and the spaces between, that network of pores through which those same roots seek out the nutrients, water, and oxygen essential to the plant's existence. Squash a soil overmuch and, like a sponge, the structure becomes compressed, reducing its capacity to hold both water and air – clay soils, with their minuscule particles and long memories, are particularly susceptible. Nobody wants to drown in a waterlogged and muddy soup, and so every plant – except for a rare few that have evolved certain adaptations to cope with having their feet constantly in the drink – appreciates a porous soil in which water finds its way in from above, but also drains freely out.

But friendship? That would suggest a degree of sentience and companionability not regularly attributed to our vegetative neighbours. Yet not only do plants form mutually beneficial relationships with soil-dwelling fungi, but evidence of far-reaching subterranean networks also shows their ability to exchange information over considerable distance. Within these mycorrhizae (from the Greek, *myco* for "fungus", *rhiza* meaning "root" – a term applied to the combination of fungus and plant) the deal goes down so: sugars produced in the green, above-ground parts of the plant are passed out through the roots into the long filaments or hyphae of the fungal body, and in turn nutrients, extracted from the soil and processed by the fungus into forms more readily accessible to the plant, travel in the opposite direction. Just one example of mutualism in nature that whispers to us of the benefits of working together, this buddying arrangement also facilitates communication between plants, where one individual, finding its leaves besieged by hungry aphids, is able to transmit a Mayday message through these pathways to allow its neighbours time to prepare a defensive chemical response against imminent attack. And so, we walk along the garden path, admiring the peace and quiet of beds and borders, while conversations take place beneath our feet.

As soil reveals to us its complexities – from a handful of dust to an elaborate blend of animal, vegetable, and mineral, home to a wealth of plant and animal life and venue for both commerce and conference – you could be forgiven for thinking that the action of plunging a sharp implement into its midst would constitute an interruption at the very least discourteous, if not outright harmful. And you'd be right… which is why it's odd that, over millennia, that's precisely what we've grown accustomed to doing.

From an early age we're taught that the thing you do with soil is dig it. We poke and prod the land to make it productive, whether in the field with ploughs, harrows, and subsoilers or in the garden with mattocks, spades, and rakes. And the accepted gardening wisdom is that the moment you think about creating a new flowerbed, a patch for your vegetables, or even renovating an existing border, you need to get digging – even better, *double digging*. It sounds like hard work, and it is. But the good news is, if you want to have a lovely garden, you really don't have to dig it.

There's no doubt that digging is hardwired into the gardening psyche, almost as an article of faith. Is it *really* possible to be good at gardening without being a fully-fledged adherent to one of its central tenets? Before we abandon the ritual entirely, it's worth spending a moment thinking over the practice of digging, if only to be sure that our low-interventionist heresies are justified by something other than shiftlessness.

I used to look forward to digging. It's a good way to keep warm in winter and, once you get going, the rhythm of the work becomes almost meditative, piercing the soil with your spade, lifting it into the air, flopping it back to the ground and then stepping to the side to begin over. Like a dance in common time,

*wcchomm * shooop * floomp * {step},*

*wcchomm * shooop * floomp * {step},*

the spade becomes an extension of your body, the rain and cold an irrelevance. In a way I miss it, though doing it properly, or double digging to a depth of two spits (a spit being the height of your spade's blade) was always a little tiresome, especially on clay. As fascinating as it was back breaking, opening a window into the layered land and, depending upon the richness of the ground, offering a glimpse into the humus-rich topsoil and all its activity, while extending into less fertile subsoil below. No matter if the strata were inverted in the process; each successive trench would be filled with well-rotted manure or compost before flopping the soil from the next on top. The theory appears sound, creating a deep layer of fluffy, porous soil, rich in organic matter, through which the roots of plants can travel unhindered. But plant roots are genetically programmed to wander, to seek out water and air and nutrients, to navigate their way around obstacles and exploit rich veins of the good stuff within the structure of the soil. Other than very straight carrots, there's little evidence that digging over the ground to create this friable, obstacle-free soil has any benefit beyond providing a workout for the gardener. After all, nature had been growing plants for millennia before gardeners appeared on the scene with their double digging. But there's more to it than that.

Digging a patch of ground year after year to the same depth can create impervious layers or "cultivation pans" within the soil, limiting the movement of air and water and restricting the ability of roots to seek out nutrients. Overworked, heavy soils are prone to smearing when wet, while the comings and goings of booted gardeners is likely, without great care, to lead to compaction, both causes of reduced porosity. Digging works counter to the soil's role as one of the planet's greatest carbon sinks, releasing stored CO_2 to the atmosphere. The coveted friable crumb texture of cultivated soil creates an increased surface area, exposing weed seeds already present to the perfect conditions for germination. Contrary to the notion that digging affords an opportunity to eradicate weeds, it exacerbates any perceived problem by chopping perennial roots into tiny sections, each capable of creating a new plant. And, as already alluded to, the mechanical disruption of the established soil ecology can scarcely be legitimized as a sustainable practice. Of course, you're going to need to open up a hole in the ground when planting, but that's an operation involving a fraction of the disruption required by regular cultivation. And so, the very good news for the time-poor, twenty-first-century gardener is that blocking out great swathes of the diary to dig over the garden, or to wander over the veg plot with a noisy rotovator, can be firmly filed under the "Really Not Necessary" column. More time to stand and stare.

If we can take from this that the soil would rather not have us furtling about within her innards, just what is it that she does want from us? The answer shouldn't be hard to come by when we consider how much the soil gives to us in terms of the plants we use to support and to beautify our existence on the planet. Looked at from this perspective, it would be ungrateful to treat the soil as anything other than a dear friend or relative, and you'd hope that none of us are in the habit of sticking sharp implements into our nearest and dearest. In fact, you'd hope that we'd be accustomed to feeding those for whom we care, and certainly to replacing anything that we take from them. If you add to this keeping them warm in the winter and cool in the heat of the summer sun, that makes for a pretty good basis for a right relationship with our soil and brings us in timely fashion to the subject of compost.

Reduce, reuse, recycle

To see a World in a Grain of Sand
And a Heaven in a Wild Flower
Hold Infinity in the palm of your hand
And Eternity in an hour
William Blake, "Auguries of Innocence" (1863)

A well-worn length of marine ply, probably a metre long and something less than a third of that in width, corners rounded with age and use, surface weathered by countless downpours and pitted from innumerable abrasions, each clumsy insult from a blade lying readily to hand patiently borne until, one day, it was discovered that the perfect tool for persuading claggy clay to relinquish its hold on rough timber was a dough scraper liberated from the kitchen drawer. It might look like a bit of old wood to you but, to me, it's the working platform from which I survey the garden, the podium that undergirds my declaration of intent upon the ground as well as the stage whereon my performance may be reviewed, though, catching sight of a genuflecting gardener, head bent and scrabbling about in the dirt, the production surely appears less theatrical than it does desperately penitent.

The board is my constant, my ground-zero and, much in the way that the space occupied by a yoga mat, no matter where you lay it down, instantly brings that intermingled sense of familiarity, ease, centredness, and calm, it feels as though there is no earthy difficulty, no tangle of weeds or incursion of over-vigorous root or questing stem with which some accommodation can't be reached. I may be kneeling, back bent and head bowed, but it's only to bring me closer to the soil. The laminated plank spreads my weight over the soft ground and prevents my knees from leaving a semi-permanent mark, from squeezing out air and moisture and compressing soil particles together beneath shallow hollows in the borders or at the edge of the lawn; just the kind of impression I'm keen to avoid leaving behind. This is cosmetic work; tickling about in the top layers of a flowerbed, teasing out tiny turves that have migrated from the lawn, disinterring the creeping underground stems of more persistent couch grass and severing field bindweed at the neck. I'm winning with the couch grass, which never descends more than a few centimetres into the soil but, with its famously deep root system, the bindweed is more of a Sisyphean task, the razor hoe in my hand *snicker snacking* through the mulch and soil to impose some temporary kind of order. There is moisture in the air and in the ground, but it's a dry day for winter, and I can walk about the garden, kneel, and flick through the surface of the beds without causing structural damage or the kind of smeary mess that occurs when working heavy, wet soils. I feel the thrill that comes from working outside in the fresh, cool air when everyone else is huddled up indoors, though some might question the appeal of spending your days with only the weather for company, kneeling in the dirt. How to explain that the earth on my knees and my hands – often on my face, wiped carelessly through my hair and inexplicably revealed as a sedimentary coda to every gratefully slurped mug of tea – is so much more than mere *dirt*. This is rich, dark stuff, full of life and, while I recognize that I garden on Wealden clay and not the dusty Mivida of Utah, I invariably experience a tightening of the gut when hearing soil referred to with such apparent disdain. Because, unless we value dirt (we don't), the use of the term to refer to our soil displays a fundamental failure not only to appreciate the beauty of nature, but an inability to grasp just how the world works. And while that's excusable in a child, for the rest of us, it's concerning.

There is more to soil than dirt. In that organic component we've already met, there is life, and death and, curiously, life again. There is transformation, mystery and not a little magic here and, while the science edifies, informs, and continues to fill in gaps in our knowledge with each passing day, there remains something ineffable in this mind-baffling combination of intermingled cycles operating on vastly different timescales that no complexity of computer modelling could ever hope to nullify; the immutability not so much of matter – which biochemistry can help us understand as minerals transfer from soil to root, to plant, animals, worm, fungus and microbe, and then back once more to soil – but of spirit, of desire, of need. We each have a need to be part of something bigger than ourselves, to understand and to experience; to know and to be known and, for me, that happens here, with my hands in the wet soil. It's a companion feeling to that comforting smallness that settles inevitably upon a person standing on the point at which the land meets the wide ocean, facing the implacable vastness of a mountain, or staring up into the boundless night sky, with the advantage that you don't have to go far from the back door to experience the sensation of holding infinity in the palm of your hand[4] exchanging Blake's grain of sand for a clod of earth. In the garden, you can find your own inspiration, your own particular reason to greet each new day with enthusiasm. For me, the appeal of circles within circles holds a fascination impossible to resist, whether in the constant rhythm of the seasons, the lifecycle of a dandelion, or the route a molecule of nitrogen will take through the soil, a plant, a mammal, into the atmosphere, and once more into the dark earth. Gardening presents us with an opportunity to get involved, always remembering that although nature doesn't need our assistance in this space, the garden, by definition, does.

Here there is alchemy, and renewal, and the chance to labour both alongside and in harmony with nature. Crucially, there is the possibility of locating meaning in our toil and, if that's too much to hope for, there is reason and purpose enough to be found in turning the compost, though this, you'd be right to point out, is beginning to sound a lot like work, and wasn't that something we were looking to cut down on? When the previous chapter offered the tantalizing prospect of laying our digging spades aside, could we not reasonably assume to have been absolved from any effort when it comes to the soil?

The answer to this depends not only upon the kind of garden we want, but also on how willing we are to let natural processes get on with what they do best. Nature can teach us a thing or two when it comes to recycling – nothing goes to waste, but instead gets broken down into its constituent parts and used again, our natural garden ecosystem running like a well-oiled machine requiring minimal external inputs in the form of fertilizers, soil conditioners, time, or money. At least, that's how it would be if we didn't get in the way with ill-judged interventions and our obsession with neatness. But there's a way to reconcile nature's way of doing business with the requirements we make on our domestic space, and the venue for this collaborative effort is the compost heap.

Left to her own devices, nature will see to the organic component of the soil by creating humus – effectively what's left over from decomposing plant and animal matter after microorganisms have rendered out the inorganic minerals and made them available for uptake by plant roots. What remains is a relatively stable presence, comprised of organic polymers, that binds to the mineral component of the soil itself. This can persist for hundreds of years, although disturbance – by cultivation, strip mining, or deforestation, for example – will cause the proportion of humus in the soil to decompose more rapidly. Humus accumulates slowly, in the absence of oxygen and, while conditions for the humification of organic matter can occur within the process of composting, humus and compost are not one and the same.

Human beings are not known for patience – the span of our lives rather dictates otherwise and explains in part our own irrepressible desire to help things along. In this, we are not alone, and can cite precedent for such expedition throughout the natural world, where specifically adapted proteins known as enzymes act to speed processes up to the point at which reactions can take place at a rate of thousands of times per second. Without enzymes, no life would be possible, their catalysis an essential function in our bodies, our plants, our soil (even, for convenience's sake, in our washing powder) and certainly in the process of both humification and composting. But whereas, despite enzyme activity, nature goes about the business of creating humus at a decidedly laid-back pace, with compost we've added our own twist, the better to hurry things along, by taking a naturally occurring activity and then, almost literally, sticking our oar in. Only in this case, the oar has been replaced by a trusty garden fork, plunged into the heart of the compost heap, and given a good wiggle around.

But I'm getting ahead of myself. Before we look at the process of composting, the question is this: assuming our garden is already a functioning growing space with a plentiful supply of that combination of rock minerals, humus, and related communities of fungi, invertebrates, and microbes we recognize as soil, why do we need to consider adding yet more organic matter? The answer involves simple arithmetic.

Looked at from one perspective, gardening is what happens when we get in the way of nature. In its most ideal form, your garden could be thought of as a closed system: plants grow, flower, set seed, and die; they fall to the floor, rot and, in the process of decomposing, return nutrients and organic matter to the soil; and the process starts over again. But each flower, or apple, or trug full of soggy, strappy leaves that we cut, harvest, burn, or send away via the municipal green waste collection, robs the garden of a little of its goodness in a very tangible way. The more we take away, the more we diminish the soil's ability to support the kind of abundant growth we all want and, one way or another, that's a deficit that needs to be made up. And very often, having opened up this idealized closed system by removing stuff from the garden, we address the shortfall by bringing more stuff back in – in the form of manures and fertilizer (organic or synthetic in origin) – an operation that, while redressing the balance of nutrients in the soil, is fit to make an economist dance and an environmentalist weep over the definition of "sustainability". All those pennies channelled into a welcome income for resource hungry products and repurposed waste; all that energy expended in processing, packaging, and transport.

And here's a conundrum. We've already established that this space isn't wilderness – this is a garden, a collaborative effort between nature and yourself, where you are allowed to grow whatever you like, to harvest bits for the kitchen, for a vase on a bookshelf, or anywhere else you see fit. You may give yourself permission to style it so that it stirs up feelings of joy and peace rather than creating chest-tightening sensations of overwhelm and angst. And so, rake and snip and sweep – surely these things are what a garden is for, so let's not give ourselves too hard a time for trying to garden in our gardens. But here's the point – it's the manner in which we go about that gardening that makes the difference – whether we blunder about in ignorance, becoming frustrated by setbacks as we try to do everything ourselves and resorting to handing over our hard-earned cash for the promise of solutions to problems we're only just discovering we have, or whether we avail ourselves of every assistance by gardening in a more collaborative way with the plants, the soil, and the air.

When we garden, then, we may as well get comfortable with the notion of putting ourselves unapologetically in the path of nature; if in doubt of the good sense of this conviction, then consider it when next weeding. Because if we're attempting to bring our own intention and purpose to bear upon a growing space, the sooner we admit this to ourselves – irrespective of our best attempts at rewilding, and our finest intentions to create a wildlife-friendly haven for insects, birds, and small mammals – the better it will be for both our sanity and our conscience. The saving grace for every child of nature who might find themselves uneasy at such an anthropocentric stance, is that the very idea of being able to exert such intention and purpose within the garden over a concerted period of time is one fit to make every dandelion, daisy, and bramble double over with laughter – an expensive endeavour in both time and resources, and one in which, it transpires, nature has far deeper pockets than you or I. But the creation of a garden is a grand thing, and there's no reason why we can't approach the process in a manner that honours the planet and principles of sustainability. Which brings us, after a brief but necessary diversion, back to the compost heap, where we take something that nature would be doing anyway if left to her own devices, relocate it to a part of the garden more conducive to our understanding of a pleasing whole (since, as we're about to see, composting is not necessarily an activity best-suited to polite society), and give it the kind of encouragement it might need, the better to align with the faster-moving passage of our own days.

The compost is rarely sited in the prettiest part of the garden, and for reasons not hard to fathom. For no matter how justly proud a keen composter might be of the crumbly, brown soil conditioner they produce, the journey towards that substance begins in a place that resembles nothing so much as a scrappy rubbish tip for vegetables and one that, since the process is continuous, never looks much better. There's at least the potential for smells, too, though a well-managed composting regime should eliminate all but the most pleasing of whiffs that hints at the activity within. For this reason, no matter how fancy a garden's compost bins are, the venue for the operation is usually well away from your favourite flowerbeds; perhaps tucked away behind a hedge or a border stocked with particularly dense and fulsome shrubs, keeping company with the bonfire pile, a stack of broken terracotta pots, and the kind of congregation of dead and nearly-dead plants that every gardener has, but few will readily admit to.

But all that is gold does not glitter, and a compost heap's raggedy appearance belies the alchemy taking place within, which is a grandiose kind of claim to make for what's essentially a rubbish dump, a place in which to corral all the bits of the garden we don't want – at least in their current form, until they can be turned into something more useful. So, in go all the weeds – though not the seedheads, as the seeds will remain viable in all but the largest (and therefore, hottest) of heaps, and not the roots of perennial weeds, from which even the smallest fragment a new plant can grow. (The first gardener I apprenticed with used to work on a "two trug" system when weeding the beds: one would be filled with relatively benign material that could go to be composted; the other was for pernicious weeds and seedheads, bound for the "black hole", which inevitably marked them out to be flung onto the bonfire, though they could as well be deeply buried or sent off to the huge municipal heaps, there to be frazzled into compliance. Now, I prefer to consign this black hole stuff to a large, lidded bucket of water, there to rot down over a period of weeks into a foul-smelling soup, which can then be added to the main compost heap with no fear of spreading weeds throughout the beds and no prospect of losing the valuable nutrients they contain from the garden.) Lawn clippings can be composted, as can the spent stems and foliage of herbaceous plants. Taking longer to decompose, autumn leaves in quantity are best managed in a separate pile, though they produce the most wonderful, crumbly soil conditioner when they do, and all that's needed are some large plastic bags with holes punched through or, better still, hessian sacks, and a place to store the bagged-up leaves for a year while they break down into leaf mould. Food waste from the kitchen can be included (avoiding meat so as not to attract scavengers) alongside cardboard such as the centres of toilet or kitchen rolls, coffee grounds, and loose-leaf tea but not, sadly, most tea bags owing to the plastic used in their construction. Anything tough or woody should be kept to a minimum and chipped or shredded before adding to the pile, and diseased plant material – for example, the leaves of roses displaying the fungal black spot that ails so many varieties – should be excluded to minimize the proliferation of pathogens.

You could wander away from a heap constructed from the above ingredients, returning in a year or so to find it diminished in size, at least some of it transformed into crumbly compost suitable for use around the garden. But in order for the composting process to work efficiently, some attention needs to be paid to the proportions in which different materials are added to the pile – a carbon to nitrogen ratio of thirty to one is the much-vaunted figure, though I've yet to meet anyone who can look at a cabbage leaf or a toilet roll tube and tell me the relative mineral content of either. Keen composters talk in terms of nitrogen-rich "greens" (green leaves, soft stems, grass clippings) and carbon-rich "browns" (wood chip, cardboard, straw), aiming for anything from an equal mix of both to one part green, three parts brown. To greens and browns, both water and air need to be factored in to prevent the pile from drying out or becoming starved of oxygen, neither of which are conducive to efficient decomposition.

All of which sounds like an almighty faff and might explain why most of us would opt to bung the lot in the bin and pop down to the garden centre to buy potting soil containing compost that someone else has made. But let me direct you to the only two things you're ever likely to need to make good compost: your nose and your eyes. The key to an efficient compost heap is, as we've suggested, balance between four elements: greens, browns, water, and air. If the compost smells funky, it's likely you have too many soft, nitrogen-rich greens; adding more brown material and giving the pile a turn with a fork or plunging a rake handle through it at intervals to aerate the mix will get things back on track. If you notice the composting process has slowed and the heap is dry, try adding water – or the noisome weed soup I mentioned earlier – to give it a kick-start, and check you have sufficient greens in the blend. Regularly turning the pile, while not essential, achieves the most even distribution not only of the initial materials, but also the decomposing agents, and is rarely anything but beneficial. It can only be a source of encouragement that, should you get this all wrong, you're still going to get compost, eventually. And if you get it right, you'll be rewarded with your wholesome, crumbly, sweet-smelling soil conditioner on a much shorter timescale, and you'll know the joy of having played a key role in its production.

The school of fling-it-all-on-a-heap-and-leave-it-alone-for-a-year-or-two clearly has the recommendation of requiring minimal effort, but what if you decide a little investment in time and effort is worth it for a superior result? With more attention to the carbon to nitrogen ratio – a disciplined and consistent mix of two parts brown to one part green – plus aerating, watering, and temperature taking, it's possible not only to speed the process up, but to produce compost with a finer texture, in which weed seeds and lingering pathogens have been killed off. Hot composting, unsurprisingly, reaches higher temperatures than cold composting and, though what you put onto the pile in either case undergoes a phased process of decomposition, the stages are more marked in a hotter pile. Each has its own characteristic community of organisms actively involved in breaking down the evidence of last year's garden into something quite unrecognizable, but entirely capable of being put to good use.

The arrival of material to the pile is like a scene from the concourse of Grand Central Station, but for microbes: a stream of new arrivals introduced with the plant matter; fresh supplies of food arriving for those already milling about. Things quickly warm up as these typically small, first-level consumers get to work; bacteria, fungi, and actinomycetes (bacteria that look like fungi, also responsible for the petrichor smell of damp earth courtesy of a substance they produce known as geosmin) going about the business of decomposition largely through chemical means. This, the mesophilic stage, is a melting pot, occurring within a temperature band between 0–40°C (32–104°F) but if conditions are right and the heap heats itself to that upper threshold, a change takes place. Between 40–60°C (104–140°F), the thermophilic stage, a new community of organisms takes over as the mesophilic biota migrate to cooler parts of the heap, towards the outer margins. This is the most efficient stage of composting, a rolling boil where microorganisms specializing in breaking down woody material and the cellulose of plant cell walls take centre stage. Like a bonfire, the consuming energy at the heart of the heap can quickly exhaust itself, and mixing or turning the pile at this point to renew the supply of fuel can help to maintain the activity. Higher temperatures spell the end for living material including pathogens and seeds, but should the upper limit creep above 65°C (149°F), the thermophilic organisms themselves die off. Once more, aerating the pile can help to moderate the upward trajectory of the mercury, keeping the compost on a vigorous simmer until all the plant material is broken down evenly and the temperature naturally begins to drop.

Cooler temperatures see the return of the mesophilic populations to the centre of the heap, this time bringing with them macroorganisms visible to the naked eye, or with a simple hand lens. Along with protozoa, nematodes, springtails, and mites, familiar invertebrates begin to lend a hand – *detritivores* – whose transformational labours are accomplished through the familiar process of eating, digestion, and excretion. Beetles, centipedes, earwigs, ants, woodlice – even worms and slugs – all play a part in the final stages as the compost matures, and the fungi and actinomycetes bring enzymes to bear upon the remaining organic matter to unbind carbon, nitrogen, and other essential nutrients, making them available for taking up by plant roots. At this stage, the presence of the gardener is required, along with shovel and wheelbarrow, to redistribute this precious black gold in generous layers upon the surface of the beds and borders from which most of its ingredients were originally removed. Though this can be done at any point in the year once the compost is ready, it's a warming task on a chilly winter's day.

And so, another cycle ends and begins again, out of sight and largely out of mind; piled in a loose heap, managed in more orderly fashion within neat bays constructed from newly milled timber or recycled wooden pallets, or tipped into purpose-made bins either to be forgotten about for months on end, or tickled and prodded with indelicate frequency for an altogether faster and more furious experience. But wherever we choose to hide our compost away, the humble prospect it presents belies a vibrant and purposeful community whose mere presence sends waves of encouragement and companionship over me, and never more so than during these quiet and solitary hours out here at the end of the year. It's partly the company, of course; the busy multitudes within the heap, working alongside me to help transform this space. More than that, more even than the satisfaction of attempting to put back what I've taken from the soil with minimal recourse to external inputs, there's something magnanimous in the garden's willingness to play along with my ham-fisted efforts to dictate what grows and goes on here. When I began to garden, I had no intention of seeking forgiveness and absolution among the beds and borders. But it seems that you and I might be in need of a little of both for the marks we make on this world and, as chance would have it, all that's required is a little time spent in humility and thought, scrabbling about on our knees in a layer of fresh, crumbly compost.

The dance

What it is to move our bodies in nature.

What it is to defy gravity, if only for a moment.

But then, what it is to come back down again. To come home, to caress with our fond presence the maternal ground where, ultimately, we'll come to rest one final time, before every speck and sliver of our myriad complexity accepts a different and no less wonderful commission.

What it is to dance through the garden, as through life.

Oh, but when I promised to tell you how to garden while doing next to nothing, you were expecting… what? A hammock? An armchair? Inactivity? I suppose we kind of covered that ground in *A garden of intention*, page 6 (you did read the introduction, didn't you?) – where I spoke of the importance to the gardener of *intention*, and how that intention could be exercised in deciding to let nature have its way with the garden while we sit back and watch. But that's not quite what I had in mind for this book. I was thinking more of the avoidance of work, the extirpation of effort which, now that I think about it, is something quite different. I was thinking of how gardening should involve as little work as possible, but as much dancing as you can muster the energy for.

Dancing is something you do, but it is not usually work. Dancing is freedom, it is surrender; it is aligning yourself with the rhythm of the earth, the sky, and the seasons, and letting that cadence guide the shape of your movement through the world. To resist the call to dance requires infinitely more energy than giving in and going with the flow, and so much of how we are taught to garden is about resistance. There is another way.

There has to be another way, because gardening is not simply a spectator sport – despite the present-day addiction to shiny screens, there is a world beyond the purely visual. But while great emphasis is placed upon how a plant looks, a flower smells, how a juicy leaf or crisp fruit tastes, even upon how the wind whispers through the tall grass, the way in which a plant feels to the touch receives scant consideration, and the same goes not only for the entire gardening environment, but for our place within it.

However it came about – whether it began in 1637 with René Descartes' declaration "I think, therefore I am" or even earlier – we have grown adept at denying our weight. Inured to the impression we leave upon the earth, we move through life in denial of the truth that actions have consequences, that our freedom to consume with impunity comes at the cost of another's freedom to exist with dignity. Until a reckoning comes, this allows so many of us to ignore our responsibilities – to our neighbours, to the world around us – but it also means we've become alienated from our surroundings, our apparent liberty being bought at the cost of our own groundedness. Unlike us, plants generally have no option but to be grounded, anchored into the soil from which they spring. Whilst growing, they are able to extend themselves – downwards for a firmer hold, and upwards to salute the sun – and to reach outwards and around, claiming space in which to unfurl leaves to catch the light, and to seek out water and nutrients below the soil's surface. Despite such distensions they remain rooted to the spot, but while the range of movement available to humankind is far greater, we are in a very real sense no less earthbound. We strive for this not to be the case, in dreams taking to the skies of our own accord, in our waking hours with the aid of aeroplanes or even through space flight – but always in the end coming back down to Earth. Nothing wrong with reaching for the stars, just as long as we remember where we're from and, in remembering, feel our connectedness to the planet. How do we do that? By making time to reach back down, deep down, into our bodies, and feeling the natural world around us; plants, soil, wind, rain, and all.

But I'm learning that, when it comes to being mindful of the space we occupy, there's something more than a kind of stationary sensory audit when we're out and about in the garden. While the practice of standing still (see *How to stand still*, page 236) – part meditation, part routine diagnostic sequence – has value in and of itself, incorporating an awareness of movement allows us to consider a more complete set of actions and behaviours that describe our own being and expression. The nascent field of ecosomatics blends a fundamental awareness of our environment with a deep consciousness of the connection between mind and body, drawing heavily not only upon deep ecology's emphasis on the value of all life, but on notions of embodiment – being fully and physically present – and a holistic approach to an understanding of the concept of the *body-mind* union. Linked explicitly to alternative therapy, if this sounds new-fangled, or New Age, it's worth bearing in mind the school of thought that stretches back in Western philosophy from the Italian Dominican theologian Thomas Aquinas (1225–1274 CE) to the Greek philosopher Aristotle (384–322 BCE) – both of whom spoke in terms of a "hylomorphic" union of body and soul – and is represented within Buddhism by the concept of *namarupa*, suggesting there's really nothing new here at all. Rather, the term "ecosomatics" is a present-day wrapper for a pattern of living that we seem to have lost along the way, reasoned ourselves out of in our post-rationalist, postmodern, post-digital-revolution existence. It's a way of being and interacting with our surroundings that, a moment's observation shows, comes naturally to a small child or to a puppy before either learn to be self-conscious, at which point the unlearning kicks in – almost certainly in the case of the human, though not always so completely with the dog. And so, we must learn it again, stripping away layer upon layer of accepted wisdom and dubious certainty passed down from one enlightened generation to the next until we find ourselves once more, happily running from tree to tree and playing about in the mud – time spent rediscovering the obvious.

A question arises as to what, if anything, could be added to such reflective practice by thinking of the way we garden in terms of a dance. Because if there's a chance of something more than a tenuous connection between two seemingly unrelated activities, it would surely deserve a moment's consideration, particularly if it were to promise benefits to the gardener (please, the *dancer*) in the form of improved mental wellbeing, healthfulness, and environmental consciousness. If the price of this kind of improvement in our daily experience is no greater than paying more attention to how we move about the garden, how we lift our limbs in space, how we feel the weight of our bodies and our groundedness to the earth beneath our feet, it's undoubtedly one worth paying because gardening and dancing are more alike than you might at first think. If you're wondering why anyone would even consider the two together in the first place, I would offer you a slightly ropey syllogism: that gardening involves movement, that movement – particularly repetitive movement – often borders upon dance, and that, should you accept the validity of each preceding premise, it's not such a great stretch to conclude that gardening could be considered a kind of dance.

It is impossible to garden for any length of time without becoming highly attuned to notions of rhythm, timing, and movement. Rhythm is seen in the passage of the seasons, one to another, in a never-ending cycle, coming ever back to the same point, and starting over. It is the beat of nature that the inhabitants of the garden – plants and mammals, birds and insects, fungi, and microbes – have no power to ignore. Timing concerns how the gardener responds in turn to the seasons, the opportunities and restrictions presented to us at any given moment to work collaboratively with the flora and fauna with which we share our space; when to cut back, to prune regeneratively or formatively, to remove a flower before it has a chance to set seed, when to gather the seed we've allowed to develop and when to sow it. When to hold things back and when to release the garden to grow, how to be present in those moments of significance that punctuate the gardening year, when to observe for instruction, and when to stare in stupefaction and awe. Movement describes actions more regular still, the range of motions we express in our bodies daily as we interact with the plants and the soil and move our bodies in this garden space. Here we can consider everything at the smallest scale, right down to the tics and flicks in the moment, those characteristic postures, fidgets, and mannerisms that allow our loved ones to recognize us from a distance purely by the way we hold ourselves. Though, naturally – and just to introduce a pleasing note of confusion lest all this seem a little too neat – movement can be drawn out over as long a period as you like, as anyone who's made a habit of, for example, photographing the same tree from one month to the next can corroborate.

There's no ostentation with this kind of dance. Neither are there any prizes for gracefulness or form – qualities for which we look to the plants, rather than the gardener. When we're freed from the tyranny of days spent at the workstation, in the contortions of the driver's seat, or nestled into the equivocal comfort of the sofa, we step outside and learn to move once more in the company of earth and sky and plants; tentatively at first, more assuredly as we gain confidence in our own gardening abilities, as we do, becoming ever more familiar with each step, each routine, every small performance of which the hours we spend in the garden are made. The garden itself places demands upon us, according always to the level of involvement we've set for ourselves, calling out of us a physical display of commitment through those devotions we habitually perform, bookended by how we enter and leave the space at the back door or garden gate, with every intervening moment an opportunity for expression. We bend and crouch and squat and kneel; we straighten, reach, take hold and pull, then lift, shake off, sort through, discard, divide and plant, and water, and mulch. And all of this, again, tomorrow, variations on a theme, our movements first, our feelings follow, and never two days quite the same. Rarely entirely still – a gardener's contemplations tend to be active, betrayed by a shifting of weight, a twitch of the nose, or a flick of the eyes – we enter a state of gardening flow, one action blending with the next, one pose becoming another, a stringing together of discrete moments into something on a scale altogether more sweeping, as everything becomes the dance. To the onlooker it might appear that we dance through the garden alone, though by now we know it's nature who calls the tune, and we take our lead from her, a *pas de deux* for those with eyes to see. And sometimes we'll dance unencumbered, but often – such accomplishment! – with props; a rake, a barrow, a pair of secateurs. To scale a ladder and dance in mid-air (though at this point any poise I might have had evaporates since nothing but the most stable of elevated platforms frees me from the uncomfortable knowledge of a large gap between my person and the ground). And still we move, negotiating space around plants, paths, buildings, pots; in and out of beds, between the greenhouse staging. And these things done in combination. And many of them performed over and over again, the diagram of our steps imprinted into the soft ground.

Over winter, we get to rehearse all this as if in an empty ballroom, before the audience arrives. With the coming of spring, the seats begin to fill; by summer, the room is packed, and we spin and bow and reach our arms about to rapturous reception from those gathered in the beds and borders. Right now, it's just you, a rake, and a robin to keep you company, and neither companion demurs as you work on your form, your extension, your reach, because now you have come to appreciate that both dancing and gardening require commitment – a disinterested dancer being about as much use to anyone as an indifferent horticulturalist. What's more, our gardening practice infused with the spirit of the dance, maybe we'll appreciate more fully the importance of warm muscles and slippery fascia, of gently stretching our back and limbs, of developing and maintaining a strong core; perhaps in this way we'll learn how better to protect our bodies from the injuries, aches, and pains attendant upon a favoured occupation. But physical wellbeing, while clearly important, can't be the end of our present explorations. Rather, we draw closer to the heart of the matter when we consider just what it is that takes the kind of repetitive movements that we undertake as gardeners and transforms them from drudgery into dance. To permit ourselves to consider gardening within the same mental space as dance – as ludicrous as it might have first appeared – confers upon us an ability to embrace both joy and delight with every weed pulled, every compost bin turned and every rose pruned, alongside a connection to the natural world that it's become too easy to forget, though our bodies are yearning to be allowed to remember.

How dull it would be only to *sow* a trayful of seeds, when every action that process involves could be imbued with consciousness, presence, and delight, our awareness of the privileges of involvement, inclusion, agency – of simply *being* – and, suffused with such energy, who could resist the call to dance? Practical considerations for your own physical welfare and that of those around you should prevent you from pirouetting about the potting shed or taking flamboyant exception to common sense considerations around bladed implements, but it would be dreariness in the extreme to deny that rush of warmth that floods into you from your gently tapping feet, upwards from the ground like the sap rising in spring, bringing with it the certainty that you are where you're meant to be, with a role to perform, and the ability to carry it off. In this way, though outwardly the gardener might appear to be trudging through their day with determined concentration – each seed settled into soil with deliberation, each pruning cut placed with precision – everything becomes the dance, your card is marked, and you begin to move in step with your partner.

That so much of what we call "gardening" happens – or could and should happen – in our heads is a truth that deserves wider recognition. Quite at what point your garden will demand your bodily, as well as your mental engagement, and to what extent you're both willing and able to commit to physical participation within that space, are matters for you and your garden to decide between yourselves. In the meantime, let gardeners everywhere embrace the spirit of the dance, and do so expansively, generously, but not – if it can be helped – too tidily.

A place for everything

It would be hard for anyone with even the slightest interest in gardening to avoid the notion of "putting the garden to bed for the winter". It sounds like something we should all be doing; certainly like something the neighbours think we should be doing. And possibly your mother-in-law. Undeniably there's industry involved in such a process but, surely, merit too? After all, the benefits of creating order are drilled into us from an early age: "A tidy room is a tidy mind". A distraction-free environment helps us to focus while the elimination of clutter brings clarity of vision. In the garden, too, when clearing away fallen foliage that might harbour pathogenic fungal spores, cleanliness might not be next to godliness, but they probably share a postcode.

But in the garden, we can be too tidy. If the space is to flourish, to be healthy and teeming with life, we need to ensure the presence of at least a little messiness, though for anyone aghast at the very idea, we could speak of incorporating areas of "controlled chaos" into their otherwise perfectly manicured space.

For tidying is a natural response to disorder and, by the time autumn hands over to winter, there's plenty of that in the garden – at least from the point of view of a person conditioned to appreciate a state of everything being "just so". Most of us, in other words. Leaves held neatly out of the way on twig and branch for the past half year or more now littering the ground, plants that filled summer borders with their bright, full-bodied presence collapsed in upon themselves, stems once proud with flowers sticking out at awkward angles, topped with exploded seed pods, flanks torn through, and bodies ripped asunder by the fervent egress of virulent progeny intent upon conquest and colonization. Then there's the mush; that soggy, slimy blanket that knits together, limited perhaps at first to discrete islands of those perennials – the hostas, the sedums – with a particular knack for drowning themselves in their own juice, later broadening out into every portion of soft, spent stem and foliage, soaked by rain and dew and broken by frost until one element is barely recognizable from its neighbour, and all now joined together into a decomposing carpet of chaos, to say nothing of the moss moving in upon concrete or stone, or the slip-sliding treachery of algae on decking.

Left to its own devices, the winter garden is a wreck of its former self. And so, we arm ourselves with pressure washers and leaf blowers to blow it all to kingdom come, shears and secateurs to cut it all down and consign it to the bin (or the compost, we're not brutes), before retreating indoors for the next few months, congratulating ourselves on a job well done.

It's a control thing, a strategy to deal with the overwhelm that follows with inevitability upon the heels of dramatic seasonal change, a milestone event in the year governed by nature's irresistible impulse to rearrange and renew and, to an extent, it works. But in tidying with such ruthless efficiency, it's worth considering how we might be depriving ourselves of opportunities to connect with our immediate surroundings and gain a deeper understanding of how the garden works in every season, settling for second best in an attempt to manage the vicissitudes of the natural world on our doorstep, when we could be enriching our daily experience every day of the year.

The impulse to tidy runs deep; that sense of the proper, experienced upon encountering an orderly workspace – perhaps a tool shed, potting bench, or kitchen – clear surfaces stretching out to the walls, tools and materials hanging upon hooks in regimented rows, or sitting neatly in allotted drawers. It's an itch that persists throughout the year, an aversion to the disorderly and unkempt that would appear to sit sharply at odds with the trend for creating wildlife-friendly gardens. You have only to stand a while within any naturalistically designed, publicly accessible garden to hear comments from fellow visitors about how "untidy" the place looks, a running commentary of confused, through disappointed to downright disapproving. Tidying, it could be argued, is what gardeners *do*, and through it, we rearrange something into a state more aligned with our own internal sense of order. Suddenly, with this admission, we have gone from a consideration of "clearing things up a bit" to the very heart of the distinction between "wilderness" and "garden". What differentiates the growth of, say, the wild strawberry edging the garden path from the same species of that plant, scrabbling along the bottom of the hedgerow (which, though also a managed environment, is about as close as many of us get to true wilderness on even a semi-regular basis)? The difference is permission; the licence we grant to plants to behave as they are programmed, to contribute towards their immediate environment in the way nature would have them do, but only to that point at which the exercise of their essential personalities begins to conflict with our sense of what is right, desirable, or useful within our own domestic space.

Tidying is saying to a plant – yes, I want that bit of you, the fresh green growth in spring, the upright turgid stems, the buds, the flowers, the fruit. The seeds? Not so much, and neither the untidy, straggly, hot mess of a thing once you've produced next season's progeny. We want you looking fresh and young, until maybe autumn, when you can entertain us with the kaleidoscopic razzle-dazzle of your annual decline. And we should be able to decide for ourselves whether we think that's ok. Almost certainly it's a response to the power of nature that's made out of fear, but if we're comfortable with the notion of domesticating animals to the extent that we can sit on the back of a powerful but acquiescent horse, or have a tame wolf sitting at our feet by the fireside, why give ourselves a hard time for wanting this simulacrum of a natural landscape outside the back door? But we're not yet – not here – talking about the process of taming the garden, the opportunity to work with plants, to partner with nature in a relationship that draws upon the strengths of both parties in equal measure. What we're looking at in this moment falls more within the realm of the clearing-up-after, the removal of the inconvenient and unsightly; if not exactly a line drawn in the sand, at least a mark scored in the soil. It's the tidying away and putting back of everything in its place, oblivious to the possibility that things might already be in their places, albeit according to guiding principles different than our own, where the appropriate station for something – a leaf, say – might have a habit of moving about from one season to the next. There's a fluidity here that we must work to grasp.

This tidying thing we do also points to a disconnect in understanding between what we might think is the purpose of having plants in our garden, and what they – could they express an opinion – might themselves think they're doing there. From the gardener's perspective, this is truer for some categories of plant than others. Home-grown fruit, vegetables and, to a slightly lesser extent, culinary herbs, are generally granted a degree of latitude when it comes to looking a little rustic and ragged, but plants grown for their ornamental value are expected to deport themselves with appreciable finesse at all times. Over the course of a year, a plant is as likely to have as many bad hair days as you or I and, while plants have a seemingly irrepressible impulse to grow (see *To raise a plant*, page 150), an inevitability must be faced; at certain points of the year, they stop; either dying, permitting bits of themselves to fall off, or starting to look something less than perky. Whether this is due to natural cycles of growth and renewal or to the action of snails, aphids, or a whole host of other garden pests and diseases, the moment at which a plant begins to betray signs of ageing is the point at which having them linger in such a state can become less of a welcome proposition, especially if décor was the motive behind their original invitation. Beginning to feel that unmistakable tingle in our fingers, we reach for the secateurs (for a quick tidy up), or the trowel (for surreptitious and speedy removal).

The plants in our gardens, we're forced to admit, are there either because we've deliberately planted them or because we've allowed them to remain, in either case expecting them to be of service to us in some manner (this, by the way, is why we become so affronted by weeds, whose very existence refuses to conform to such an anthropocentric model and hints at perspectives other than our own). But while philosophers through the ages have wrestled with questions of human ontology, the existential imperatives operating within the plant kingdom seem refreshingly straightforward – to live, to flourish, to reproduce and, when all is done, to die in such a way as to benefit the community. This might involve, at the appropriate moment, making oneself attractive to pollinators, or wrapping one's seed in packages contrived to tempt those who might disperse them, but not really to put on a year-round show of immaculate poise for you or for me. In this, we come to understand that harmony in the garden occurs in those moments of alignment between our priorities and those of the plants around our home, autumn and winter bringing with them weeks of prolonged dissonance as not only the plants, but the elements, too, seem hell-bent on reminding us that, out here, we're not in charge.

Which is all very well, but there are practical matters to be considered. Even those of us who delight in the annual spectacle of decay and decline being played out each winter need access to the soil; to plant bulbs, and blanket with manures and mulches. There's already a layer of decomposing plant material to nourish the soil and raise a community of plants, but these aren't the ones we generally want to grow. And so, with thanks to nature for the effort, we rake and snip, gathering this material up and carting it off to the compost heap, before adding very much the same kind of thing in its place and this, though it might seem crazy, is something we need to give ourselves permission to do. Here is an illustration of that partnership we have entered into; we know full well that this ground knows how to grow plants and tend its soil and feed its resident creatures but – not unlike a child insisting on helping out with the baking, and getting quite particular about how it should be done – we want to help, to feel useful, to have our say and in doing so, to feel involved. Somehow, it's not quite enough to accept nature's generously given gifts – we want, in short, agency. And so we acknowledge that, while it's a fine thing to forage, it's great to be a gardener.

I would like to be able to manoeuvre my wheelbarrow through the borders now there's so much to be gathered for transport to other parts of the garden and, once deep within the beds, I want to have a good look at the plants – particularly the herbaceous perennials that, without a woody structure to support them, are so radically transformed by the passage of the seasons. I want to tuck them up for the winter in a thick blanket of mulch and, as all their activity slows, stills, and disappears back below the ground, I want to introduce a marker – a 25 cm (10 in) length of bamboo or hazel, perhaps – so I'll know where to avoid shovelling in bulbs over winter, or excavating holes for the extra hellebores I'll be unable to resist buying in spring. I need to make sense of the chaos before me, quite the opposite of that continuous, billowing tapestry I've been looking to create throughout the rest of the year. What my gardening brain needs now is to strip every planting combination back to its constituent parts, and I can't do any of this entirely unnecessary necessary activity without first having removed the detritus of last year's garden party. Pulsing to a seasonal rhythm, the garden breathes out and in, and just now everything out here contracts and pulls back as we ride the most significant inhalation of the year. If ever there was a moment to celebrate the spaces in between it's winter, and I'm lending a hand by creating a little extra definition. I need discernible borders and edges, not only to the landscaping elements – paths, lawns, patios, beds – but to the individual clumps of plants, too.

There are times when it feels as though I am carving into the garden, tinkering decoratively with the topmost layers, and the alterations I have in mind just now reach down no deeper than this. My chief zone of operation exists close to the surface of the soil, perhaps a few fingers' width above, where the spent remains of leaf and stem cling stubbornly to their former home, though there's no life for them there now – at least none resembling one they've known most recently and not, if they're in my way, if I have anything to do with it. Sliding my board into any gap I find between plants, I make my way through the bed on my knees, clearing paths through and around slumbering perennials, scraping the husks of once proud and perky foliage into a trug, there to be joined by weeds and annuals wrenched whole from the ground, and any stem too bedraggled to comport itself with the dignity appropriate for the winter display and the buffet laid on for the wildlife. I clear some areas and ignore others, leaving an eccentric kind of paisley in my wake, its pattern governed by half-formed plans of imminent bulb plantings and proposed sorties with the wheelbarrow. Whether or not the centremost stems in a clump are given leave to stand, neither languid intrusion nor floppy incursions into the spaces between shall be allowed.

When it comes to the separation of one part of a plant from another, nature tends not to cut cleanly (the exception being leaf fall in woody plants, but you'll have to wait until *What goes up*, see page 266, until we look at that), permitting the nibbling activity of small creatures and the bumbling progress of larger ones, the ripping of the wind, the snapping of limbs under load, and all the jagged-margined wounds such calamities leave behind. Each of these actions result in just the kind of complex, frayed, and multi-layered edge that horticulturalists are taught to take every measure to avoid when slicing through plant tissue, the better to guard against infection. Nature takes more of a big picture view, one where the emphasis is always upon life, and as much of it as possible, where the wellbeing of a single individual plant might well be offered up against the prospect of rich communities of bacteria, other microorganisms and fungi, and the dependent populations from higher trophic levels, which such a smorgasbord would come to support over time. Nature knows what she's doing but, since her interest in the diversity of existence is dwarfed only by the degree of her indifference to the amount of money I've spent on plants for the garden this past year, where she nibbles, snaps, and rips, I snip, a pair of secateurs never far from my hand, blades sharpened to a cruel edge, springs taut and all parts gently glinting with an oily iridescence. A well-designed tool needs to feel like an extension of the limb that wields it, the edge of its blade or the tips of its tines continually employed as the point of contact with the work piece, an interface between maker and made. So connected do I feel to my secateurs, it remains a source of great consternation that, upon extending my arm, they don't spontaneously materialize in my hand. It's something I'm determined to work upon but, until such manifestations transpire, I keep these hand shears close, holstered on or about my right hip when not actively employed, but always ready at a moment's notice. The understanding between gardener and gunslinger runs deep, and the instant someone starts making pearl-handled secateurs, I'll be the first in the queue.

But we're clearing today rather than pruning, and my secateurs could be excused for feeling this work beneath them; chopping through dead plant material occupying a position barely a few degrees above that of indiscriminate hacking, with no call for that degree of precision required when cutting into living tissue. They snap and gnash their blades together all the same, hungry for stems to grip and bite and slice, and while hedging shears would cut through three times as much in half the time, there's no finesse in such over-ardent chomping, and I appreciate the control a smaller pair of garden scissors can offer. For, all the while, I'm mindful that tidying can be taken too far; for that which one hand reaches to grasp while the other aims to sever could be someone's home, or someone's dinner, and if not yet just now, then potentially so in the very near future. It might be a matter of utility for me – how best to get about the garden without, for example, going head over heels on a patch of slippery leaves on the path, how to think about the behaviours and needs of one plant over those of its neighbour – but that which is expedient for us is not necessarily useful for the other garden residents. What makes for clean lines and straightforward access, particularly at this time of year, could easily begin to intrude upon sterility. And I want my garden full of life even here and now in winter, whether foraging for energy-rich leftovers among the slim pickings of a denuded landscape, or snoozing quietly away in hollow stems and vacant seed capsules.

For this reason, on that day when we are held to account for our gardening behaviour, if such an hour ever comes to pass, I will stand up and be counted with those who would err on the side of the messy (a thing that will come as no surprise to anyone who's seen my garden), though I'll be forced at the last to make a confession. For all I would champion the value to wildlife of the garden's winter clutter, for all the advocacy of discernment in selecting what may be cut and what should be left behind, I must always in the moment have my wits about me or, secateurs in hand and trug by my side, a mist descends. As the impulse to tidy takes a hold, I cut and I rake and I gather away till all is prim and neat and proper, and I am full of shame. Bare earth and bald crowns – a worthless horticultural aspiration with surprisingly high currency. It's a behaviour that has held me in good stead as a jobbing gardener, when the aim is as much to restore for the householder a sense of being in control – "*Look on my Works, ye Mighty, and despair!*"[5] – as it is to elevate the priorities of their very nearest and most vulnerable neighbours. But it's a poor and joyless way to garden – and "*the lone and level sands stretch far away*"[6].

Again, I seek the middle route, between nature's desire to let everything rot where it falls, and my own apparently deep-seated instinct to tidy it all away. I can, I think, bring myself to do just enough to carve my paths through the chaos, but I'll be leaving some standing, for the birds and the insects, of course, but also for the frosty display. The winter garden sparkles, but so much more so when you leave some of the scaffolding in place.

How to PLANT A BULB

Brook no distractions when unpacking your bulbs. Some activities warrant a degree of ceremony, and being bookended by a breath. To which end…

Breathe in, hold for a moment then, slowly and with some noise, breathe out.

Open the bag; by rights it should be labelled "Hope & Promise", but something like "Tulips" or "Lilies" is perhaps more likely and, really, it's all the same. Pour the contents out into a small pile; taking up an individual between thumb and forefinger, apply firm but gentle pressure. Within a dry and crinkled paper coat there should be taut, plump resistance; discard the squashy, the damp, or – heaven forbid – the mouldy.

Run your thumb over the rough, whiskery basal plate from where the roots will emerge, and continue turning until you get to the point that marks "this way up"; a gentle peak now covered by paper-thin skin and the fleshy, nascent leaves through which the emerging stem will shortly push its way. A plant instinctively knows where to find the sky but burns energy in righting itself at the expense of flowers.

Breathe in, hold for a moment, then slowly out.

You are ready to plant. Come spring, flowers will announce that the soil has forgiven you for repeatedly puncturing its surface, but the invasion should be acknowledged, and permission sought, before clumsily inserting your chosen few into a complex and functioning society. Select your hole-making instrument based upon bulb size and ground conditions. An ordinary trowel will do for tulips in a sandy soil, but for clay you'll be better served by a bulb planter that will leave a neat hole with every plunge. A thin trowel

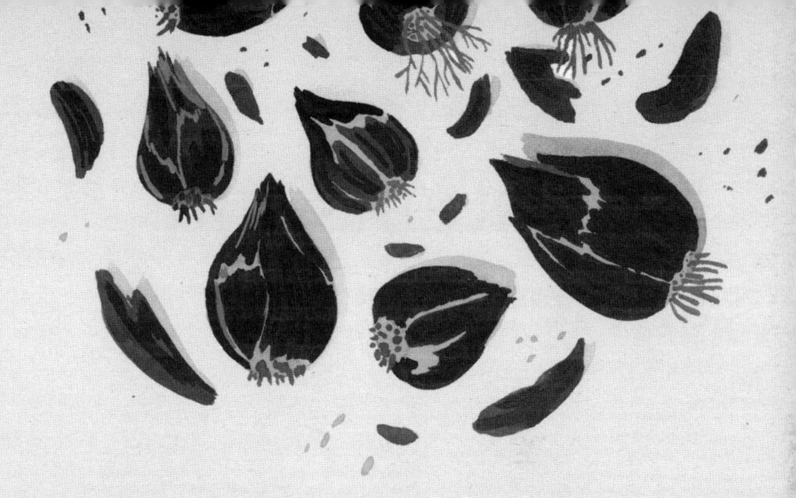

with a point or a dandelion weeder will be ideal for smaller bulbs – muscari, or crocus, say. Whoever invented the long-handled planter with treads forgot you must still bend down to place the bulb into the hole.

Find a board on which to kneel – the ground will be soft, and the direct application of a gardener's knee will leave a longer-lasting impression than strictly desired.

A breath. This next bit requires a little visualization. *As before, in, hold…*

Picture the scene with the bulbs you're about to plant in bloom… *breathe out with the noise of the wind in long grass…*

…and scatter the bulbs where you see flowers. Avoid the much-trumped technique of throwing the bulbs into the air and planting them where they land; bulbs are sociable individuals and invariably roll into tight groups.

Make a hole to a depth two to three times greater than the height of your bulb, and drop your bulb neatly in, pointy-end up, before covering over with the soil previously excavated, and firming it gently back down into place. Repeat, until every bulb is tucked away. In a month or two, your eyes will thank you for the effort you have spent today.

Stand up, step back, and stretch well. Your body will thank you for this tomorrow.

Bare stems and skeletons

Learning to embrace a looser look shouldn't bar us from celebrating some of the austerities attendant upon the garden in winter. A magical time of bare stems, of silhouettes and skeletons, a season when our gaze, unhindered by the leaves that at other times of year give body and mass to the planting, can move searchingly and critically around the space. It's a moment when form and placement are thrown into sharp relief, when we're challenged to leave the plants to grow away according to their own design, or to get stuck in and engage with the natural processes that will shape the garden throughout the coming year.

As milestones in the horticultural calendar go, "Betwixtmas" has yet to receive the recognition it deserves. But there comes a point towards the end of December, somewhere within that soporific lull that descends after Christmas and only begins to lift with the welcoming in of the New Year, when many of us begin to crave simplicity, to have everything pared back. Maybe it's the inevitable reaction to groaning tables of festive food, the richness, the salt and fat, the sour, the sweet, and we just want, what... salad? Well maybe not quite salad, unless it's a superlatively good one, but definitely, less. A clear soup, some wilted greens, a simple pilaf. We admit to ourselves the emptiness of excess and resolve never to look upon another mince pie until next November before, turning our backs on tinsel, we head out into the garden for a cleansing course of damp soil, cool breeze, and birdsong.

The winter solstice is the true turning point at this end of the year, and by now, the rest of the natural world has had over a week to respond to the lengthening of days, while our social calendars and obligations have kept us busy indoors. For the gardener, the end of this short, self-imposed exile is restorative, and perhaps the garden feels it too. It will soon, as this is a time where considered appraisal leads in short measure to action.

By now, all the bulbs are in the ground, or as good as, since it's still cold and there's time yet to get to the box of reduced-price tulips that I stashed in the shed a fortnight ago. Now, such hectic impulse shopping feels like the relic of a former life, and my newfound puritanism responds to the bleak stillness beyond the back door – with snow and frost, this is the landscape of Christmas cards and documentaries on the Northern Lights, and even under a grim, grey sky with droplets of mist falling fatly from every twig and stem, there's a certain dour beauty all around. The garden reveals itself in winter, so much so that, casting a gaze across the space and allowing my attention to drift from one plant to the next, each divested of all finery and offering up its nakedness for inspection, it's hard not to be struck by a mingled sense of vulnerability and wonder. Is this really all that's holding up those blowsy and billowing borders of summer? But more than that, now that I can see through this shrub, into that corner, clear across these plants that only a few weeks ago were drawn like a veil across that section of the garden, the whole space seems so much bigger. It's an optical illusion, but one to be taken advantage of – winter lavishing upon me room to move and think. Clarity is out here for the taking, though I might have to do a little pruning away of the old and unwanted before I can get to it.

Now, more than ever, a good path is key, its role as the spine of the garden never more candidly exposed. The garden path dictates how we move about, how we orient ourselves within the space, navigate between its borders and gain access to the planting; it represents the point from which we depart, plunging headlong into the beds, and to which, scratched and muddied and bearing armfuls of clippings, we return. Of such central importance to the garden is its main path that, in those cases where, for whatever reason no formal provision for such a thing has yet been made, it will manifest of its own accord, a desire trail appearing gradually as an increasingly well-trod route across the lawn – though anyone who's spent more than a few winters in the company of a grassy thoroughfare will find themselves yearning for one of brick or stone or pavers to save them from the inevitable quagmire. (Similarly, a poorly-routed path, no matter how impressively landscaped, will inevitably be superseded by the more direct route unless care is taken to obscure with voluminous planting sightlines to the various garden destinations – the shed, garage, clothes drier. In the company of all animals and birds – and especially crows, who are famous for it – we like to take the shortest route between two points.)

More than this, the path represents a physical manifestation of the gardener's intent, leading us away from the closeted safety of the house into the unknown wilderness, a declaration of purpose and a commitment to engage with the natural world beyond the doorstep. It is at the same time a known quantity, barely changing from one season to the next, a place of relative safety from which progressive sorties into the unknown on either side can be mounted, surrounded not by enemy territory but fertile ground brimming with possibilities, the promise of deep, sustaining relationship and a sense of fulfilment. Day by day we advance and retreat, the garden in a continual state of flux, the gardener growing older, and the path remains constant.

The path, just now, needs a bit of attention.

Barely discernible beneath a thick blanket of snow that arrived while we slept, the notion of preserving the pristine surface persists for as long as it takes to brew a pot of tea, at which point the prospect of plunging booted feet into thick, virgin powder becomes too much to resist. There are snowballs to be thrown and snowmen to be made, our feet locating the path below through muscle memory or some innate sense of spatial awareness we're not even conscious of tapping into. By lunchtime, a coal-eyed Frosty winks from beneath his jauntily angled hat upon a garden that's received more trampling and scraping in the space of a few hours than it usually sees over the course of a month. Fluffy snow compressed by a zigzag of boot treads turns to ice along the path, falling temperatures consolidating the glinting veneer into a truly treacherous surface. It's bad enough over grass, but on brick or paving it's downright dangerous. To restore safe passage, shovelling ensues; perhaps there will be a scattering of sand, augmented by a very little rock salt (the plants in your garden, like the food on your plate, will not thank you for being over-seasoned).

Snow calls a halt to much of the work in the garden, though really, it's frost that poses the greater risk to plant tissue, and many a "milder" snow day has seen me out pruning apple trees and roses as my fingers turn blue from the cold. But the invitation to experience the garden under a heavy fall of snow should never be declined, if only for the opportunity to seek out the tracks of birds and animals, and gain an appreciation of the paths that other use when navigating around the space they share with us. Snow also has the benefit of wiping out the smaller details, leaving only the larger forms and volumes in place – a temporary and partial wiping clean of the slate, which brings with it a venue for daydreams of *what ifs* and *maybes* to flourish unhindered by the restrictions imposed by what's already in the ground.

Here, whole winters may pass with barely a flurry of snow and, despite the disruption it brings, the arrival of a good fall that settles on the ground, transforming our surroundings and intensifying, through a billion tiny reflecting lenses, the available light on these shortest of days, is still sufficiently rare to be met with rapt enthusiasm. Frost, though, is a more frequent visitor, the first footfall of any winter's day far more likely to sound as a *crunch* than a *flump*. But, for all the familiarity, frost has its own magic, too, reserving its deepest enchantments as reward for past forbearance and choosing, as subjects for the greatest transmogrification, those stems and seedheads left standing.

We understand their value to wildlife and recognize the rich and toothsome textures they add to the winter border, as well as the importance of vertical interest when so much else has either collapsed or been removed. But our appreciation of winter's hold upon this space is incomplete until we step out into a frosty display by the blue light of dawn; spider silk weighed down with a diamond crust, the cool and crystal glint of ice-rimed edges catching the watery yellow sun – a midwinter tableau more beautiful than bleak, and another invitation to be present without feeling the need to do anything other than be part of the scene.

But sooner or later snow thaws and frost melts away, restoring the detail and leaving slush and mud in its wake. Those with the fortune to have been visited by the lucid perception that icy cold can bring need now to hold on to such insight among the damp brown noise and litter of the winter garden. If – as is quite normal for these parts – such a clarifying whiteout remains a distant rumour, then we must work at creating this mental space for ourselves. For me at least, that begins with clearing my way.

Today, there are leaves, but not for long. The illusion of order can quickly be established by the simple expedient of clearing the paths – as though the prospect of purposeful and clearly laid-out highways through the wilderness presents a powerful visual metaphor which the brain is more than content to accept. It is the very first thing I do when faced with the kind of chaotic garden scene that, at this end of the year, abounds. Riot and collapse rule upon either side, but ahead and behind, there is no such ambiguity, and the skeleton of the garden's various routes, freed from obfuscating detail, begins to reflect the bare bones of the trees and shrubs that rake the winter sky. If I'm feeling fanciful, I imagine that I'm unclogging the arteries of this semi-natural space, clearing a passage along which the garden's *chi* can course unhindered: it seems to be doing the trick in my own mind. However it works, leaves swept or blown aside into the beds and borders (they can be dealt with later, if I decide I don't want them there), everything seems more manageable, and whatever shape I envisage for my winter gardening – whether

I'm intending to plunge in and engage with pruning saw and secateurs, or opt for a less hands-on approach of mindful daily presence, sensible of lessons to be learned from careful observation and thought – this straightforward but necessary step has transformed my festivity-fugged head and set the stage for a new year of gardening.

More often than not, the impulse to get involved is too great to resist, and I do engage, bringing saw and loppers to bear upon slumbering wood. Having cleared a runway through the space from which to survey the plants, I take a cue from those herbaceous perennials who, for their contributions to wildlife and winter structure, were so recently spared from the chop and spend a good while standing about amidst it all. With a proportion of the garden's population now bereft of leaves, it's easier to pick out an individual from the crowd, assess the form of a single plant, trace the growth of branch and stem, admire the shapely and identify the congested or the tangled. To any onlooker raised within the school of garden busyness, these minutes spent gazing in rapt attention toward the borders must seem time ill spent, but that which you don't see is as important as that which you do, and the matter of deciding where and what to prune must be weighed with care. An invisible bubble surrounds each plant in your garden, and how far you allow one plant to intrude upon the envelope of its neighbour is a matter of discernment; a balance of space and not space towards which your garden would have you take a considered approach.

In summer, the purple smoke bush appears from a distance to be wreathed in mist or smoke, its spectacle created by plumes of tiny flowers. Peering over the lid of my laptop, I can see that last year's inattention has allowed mine to impose itself without the barest nod towards social distancing upon the territory of the long-suffering amelanchier with which it shares a bed. Certain shrubs are supremely cooperative with the gardener's desire to prune them in winter, and the smoke bush, being among their number, acknowledges the compliment later in the year with a show of larger than normal leaves. These come at the cost of its eponymous floral effect, since the flower buds form on woody stems more than a year old. Not everything in the garden responds with such accommodating grace, and many a long-suffering shrub will fall victim to an indiscriminate hack between now and the blooming of the tulips, hostage to the idea that a plant, having outgrown its allotted space, requires putting back in its box. When the overriding consideration for pruning a tree or shrub is a reduction in size, the result is unlikely to be either aesthetically pleasing or horticulturally satisfactory, scant consideration given to the plant's form, the position of its various limbs or the preservation of the kind of open structure that allows the passage of the breeze, bearing away upon its cleansing gusts those fungal spores that might otherwise linger with injurious intent.

Perhaps more even than an excuse, we have an obligation to stand and stare as a precursor to making the first cut; before the selection of this limb or that, there are decisions to be made over what stays or goes. When it comes to the shape of a stem or branch as it describes its journey through the air, I find myself drawn to the twisted and leaning and, perhaps because there is more room to wonder about the cause of such contortions than when faced with the straight and true, I'm strongly inclined to give such gnarly character leave to remain. Admittedly, there is energy and intent in the kind of purposeful growth that shoots straight for the sky without turning to one side or the other, but such laser-focus can be unnerving, and I'm keen to remove anything that flaunts such forthright tendencies. But it wouldn't be gardening if there weren't a conundrum, and winter pruning of deciduous trees and shrubs, particularly when indulged in without sufficient restraint, encourages precisely the kind of vigorous, upright growth I find myself wanting to dispense with. Why should this be? For an answer, it helps – as so often proves to be the case – to think like a plant.

Autumn turns to winter, and you – a woody perennial of approximately 4 metres (14 ft) in height – fall into a contented sleep, having prudently enjoyed a meal that will provide all the energy required not merely for your hibernation, but for the initial growth spurt that follows. Being the unwitting victim of a significant process of editing upon your person while in a dormant state, you awake in spring to discover that you're now a shrub only 2.5 metres (8 ft) tall, but with the calorific resource appropriate to a much larger plant – the plant you were when you shut down for winter. All that energy requires an outlet. Rather than simply existing as a freakish stub of your former self and fountaining fresh sap from cut fingers, you call upon every latent growth bud that slumbers beneath the bark to grow wood as fast and straight as can reasonably be accomplished, transforming you into the woody, witchy thing that haunts the landscapes of the darkest fairy tales. These long banshee fingers – water shoots from the epicormic growth described – become a feature of the winter scenery, as the bare bones of every tree are revealed in silhouette against the sky. To anyone who simply wants to restrict the size of an unruly shrub, it can seem as though a plant is intent upon resistance, a response that can be tempered by staggering major reductions over several years, removing no more than a third of branches in any one season, and by pruning in late summer before leaves surrender up their energy for storage over the winter. But it's a biological response in woody perennial plants that has been harnessed by humans for millennia, providing straight wooden poles of ash or hazel for construction, firewood, and – in the garden – for plant supports, with more pliable wands of willow for the weaving of basketware. The closely related processes of coppicing (cutting trees to a short "stool" at, or near ground level) or pollarding (pruning growth back to shortened trunk or branch) are as central respectively to the management of woodland and of city street trees as they are to horticulture, where the youthful vigour of the juvenile stems that burst upwards from each stump bring such character to the winter borders. With careful management and understanding, such energetic rejoinders to our attention prove desirable in terms of both their economic and aesthetic value, though their appearance to the uninitiated can be a source of alarm and frustration.

For the duration of this freewheeling pause at the top of the year – these in between days merging one into the next in a fog of fir, clove, and brandy – we turn to the garden for what it offers our weary souls in terms of clarity, escape, and realignment. Right now, it might seem a one-sided transaction but, after the fashion of all good relationships, it will soon be our turn to carry, to soothe and to tend, and the garden doesn't seem to be keeping score. The world around us is making resolutions for the new year while these few still, bare, quiet days arrive like a Christmas gift we'd somehow overlooked in all the hustle of the day itself, coin to be spent in reorienting ourselves, and committing to the garden over the year to come, looking at once inwards and forwards, but not forgetting in all our planning and pruning and anticipation of future growth to look around and delight in the here and now. Because, now we've been out here a while, it seems the stems we once thought bare are not so; there is colour with jewel bright berries, garnet haws, and scarlet hips – winter warmers for the birds. There are the brilliant, straight branches of pollarded willows and coppiced dogwoods, cut back hard every spring to fend off the dullness of old age that afflicts us all and burning now with all the fire and enthusiasm of youth in shades of red and orange and yellow. And the scent – there is perfume, too, from flowers too eager to wait for leaves to clothe the branches, and the air is spiced with wintersweet and witch hazel, winter-flowering honeysuckle, and *Viburnum* x *bodnantense* 'Dawn', these two last two seeming to compete over the prize for gnarliest stem with most fragrant blossom. For all the spartan restraint of the Betwixtmas garden, there is treasure here among its bare stems and skeletons.

Ivy, bramble, and briar

According to their peculiar life cycles, most plants occupy the garden spotlight for only those few weeks when they are in flower. It's the most attention-seeking of evolutionary mechanisms, designed both to exploit the availability of pollinators and outdo the competition and, once the deed is done, they can fade once more into the background and get on with the business of growing. But some characters, by sheer persistence and force of personality, and careless of blossom or time of year, remain very much at the forefront of the gardener's experience, either taking full advantage of a clear field while so much around lies sleeping or, in the absence of leafy camouflage, being thrown into sharp, and often painful relief. But grumble and curse as I may, hauling upon yet another long tangle of creeping stem or dragging several metres of prickly cat o' nine tails down from the canopy of a tree, I can't deny the glow of admiration I feel when faced with such irrepressible spirit. And if that glow won't be sufficient to keep me warm out here on these cold, damp, winter days, then battling with ivy, bramble, and briar will.

As a coterie of plants, the ginseng family Araliaceae are nothing if not tropical. Many with lush, waxy foliage, some – such as the false castor oil plant (*Fatsia japonica*) or the enormous rice-paper plant (*Tetrapanax papyrifer*) – extending upon the end of long stems large, palm-shaped leaves with an open-handed shrug; *you're right, I'm huge – what you gonna do?* Others we know from our homes and offices, where umbrella plants (*Schefflera actinophylla*) bring a touch of the leafy exotic to our interior space, while the Devil's walking stick (*Aralia spinosa*) strides through the landscape and into the gardens of the south-eastern US, bringing a touch of menace with its spiny, unbranched stems.

Who would have thought that a plant so central to both rural and urban landscapes that we scarcely remark upon its presence would emerge from such an extrovert family? And yet these are the relatives of our common or English ivy (*Hedera helix*), though upon closer inspection, this ubiquitous vine – a more complex and intriguing presence than casual acquaintance would suggest – is far from diffident itself (indeed, something of a terror across the Atlantic, where its prodigious growth can threaten indigenous flora). Not so much an inveigling thing, neither surreptitious nor sneaky, ivy operates while hiding in plain sight, relying on its familiarity and our inattention to go about its business unchallenged. Quick to move in upon the forgotten, abandoned, and unloved, and entirely honest about its territorial aspirations, ivy extends its long, branching stems in any orientation, effective both as a climber and as ground cover. An enveloping, evergreen blanket relentlessly advances at a rate of up to 2.5 metres (8 ft) each year and, though new plants will take a couple of growing seasons to establish, each may endure unmolested to span four decades. Perhaps due to its reputation as a portent of death and decay, the presence of ivy – once registered – is at best tolerated and, more likely, feared. "You want to get rid of that ivy," is a phrase intoned by dads everywhere (it's always dads, it must be in the handbook), mindful of imaginary destruction wrought upon the fabric of your house by its presence. But while any brick wall might prefer to be ivy-free, due to that plant's particular properties, prevention is preferable to cure.

There are different kinds of climbing plant. Some that clamber about loosely haul themselves over anything in their path with the aid of prickles, but require being tied to a support for any disciplined effect. Others form attachments by twining their leaves or stems around convenient objects and hanging on for dear life. Still more require no support, fastening themselves to surfaces by a combination of friction and stickiness, and it's this impressive self-clinging superpower that gives ivy its bad rep. Behind an apparently miraculous ability of adhesion are the soft, adventitious roots that can grow from every swollen leaf node along the plant's stem. These winkle their way into any small crack on the supporting surface, grappling hold by means of tiny root hairs and then transforming (through a cell strengthening process known as lignification) into woody anchors. Just to make certain of their grip, the roots then exude a kind of plant-based superglue which ensures that, tug as you may, they're not going anywhere. Almost certainly the worst thing you can do for your masonry is to pull a mature ivy plant away from brickwork, the futility of persuading the plant to loosen its clutch rendering the feared outcome far more likely – at best, the stem will come away and leave the roots in place as an unsightly blemish, though just as likely sections of flaking mortar will be pulled loose, bumping "repointing the brickwork" far higher up your list of priorities than if you'd let the ivy be, a wildlife-friendly, insulating blanket for the house.

Given free rein, ivy will – in time – contribute to the decline of any tree up which it climbs. But the sorry practice of indiscriminately cutting away the thick, fibrous stems of mature ivies results in the loss of an important wildlife habitat, while invariably leaving the upper portions of the plant orphaned and dying, hideously marooned high in the tree's canopy. The initial stages of an ivy's journey up a tree are quite benign, even ornamental; the familiar three-lobed leaf of the juvenile lower portion of the plant (because ivy is also a changeling, transforming its appearance as it matures and gains height) in glaucous shades with white, silver, or cream variegations complementing the bark to which the red-brown stems cling. And though woody vines (or lianas) like ivy draw sustenance from the same soil as the trees upon which they depend for support, you'd hardly call it competition. Neither do they parasitize the tree by invading its vascular system; since they're rooted at ground level, it matters little to such vines whether what they're clambering over is dead or alive. But once ivy reaches into the tree's canopy and advances its own mature foliage, there is heavy rivalry with the host over sunlight. It's evident, at a glance, which party is flourishing, and which is suffering from the arrangement, and so it seems fair to conclude that ivy is bad for trees – added to which, the weight of the ivy's flourishing leafy growth can make an ailing tree top-heavy, whilst acting as a sail to catch every potentially destructive gust of the cold winter wind. But, looking at the garden as an ecosystem – in terms of the quantity of biodiversity it's able to support – it would be hard not to play the numbers, sacrifice the tree (which will still, as standing dead wood, provide homes to countless invertebrates and fungi, as well as food and nesting opportunities for birds and small mammals), and encourage the rich habitat and nectar reserve of the ivy.

This is more of a long game than we gardeners are accustomed to playing, aside from which, we grow fond of our trees as individuals, a relationship which should allow for close (if only, perhaps, one-way) observation. So while the presence of a mature ivy is unlikely to benefit an individual tree over the long term, it shouldn't be beyond the resources of the gardener to maintain any bark-bound specimen in its juvenile state, before it can become an overly muscular presence. I've noticed that the fat triangular blade end of my secateurs, slipped between tree bark and a young ivy stem, is an invaluable aid in persuading the climber to relinquish its hold and, purposefully levering a short section away from the apple I'm currently in the process of pruning (10 cm/4 in is about standard before the root hairs get wise to you and the section snaps), I can't help but wonder if the ivy knows something I don't; if it's actively seeking out trees already in a state of decline in order to support its clamber towards the sun. At a time when we're learning more about the biochemical signals sent between individual forest trees, it seems a small step to suggest that an ailing tree's weakness might be sensed by other plants through some imbalance in the surrounding ecology, some alteration in the local population of microbes and fungi indicating the onset of senescence and extending an invitation for opportunism.

Ivy only flowers on its mature sections, the dull yellow geometry of its domed and branched blooms appearing from amongst glossy green, heart-shaped leaves, providing a feast for pollinating insects in autumn just as everything else in the garden has gone to seed, a favourite food stop for bumblebee queens preparing themselves for the stress of hibernation. Creeping across the ground, thin stems wander into the back of borders and wind themselves about, weaving a densely-foliated mat that insulates the soil from hard frosts, providing habitat for invertebrates and cover for foraging birds, and there's a judgement call to be made over whether it stays or goes. Today, half of every wheelbarrow load is a tangle of ivy, though I leave the calorie-rich berries – in their smart, jackdaw livery of charcoal grey with black caps – for the birds. Do I leave as much behind as I take? I've not done the calculation, but that feels about right, for ivy is as much a part of this garden as I am, with deeper roots in the surrounding landscape and, even if I could banish it, why would I want to?

Some things you have to grapple with. It's a matter of boundaries, a question of respect or, like Jacob with the angel, a demand for recognition. When you decide to garden with, rather than against nature, it's like any partnership – there's a period where each of you sizes the other up and, inevitably, there will be moments where one might try to slip something past unnoticed. Nature will always have the upper hand; you must decide if you're simply going to let her roll over you, or if you intend to do something to earn her regard for the fleeting spell – only a blink by her reckoning – during which you'll be shaping this space together. You can choose how you'll make a stand; go for something creative, and collaborative, rather than taking a futile tilt at subjugation, but at the same time, don't take any prisoners.

When nature sends these interlopers scrabbling over your walls and laying claim to all they meet, it's time to push back.

Brambles respond to no polite hint; no sidling up to mutter tactfully that their presence is no longer desirable. Forcible ejection from the establishment is required, with all the disruption and kerfuffle that might suggest, though this is easier said than done with a prickly guest who might look slight and wiry but who, it transpires, is sporting heavy boots. You may view the fierce prickles as your main challenge and, hands encased in thick, leather gloves, feel adequately equipped for the task before you, but do not be fooled. The gardener who emerged victorious from a tug-of-war with a bramble is yet to be born. A friend and colleague, younger and fitter than me, ended up flat on his back for weeks, that back having emitted the kind of "ping" that no back was designed to emit during an encounter with a bramble that took exception to being evicted. Where you see sinuous green stems, think high tensile steel cables, tied down to a lump of cast iron, buried in the ground. Pull gently, locate the roots, and go lay your hands on a mattock to dig the beast out.

And yet, aside from failure or success in the more manicured areas of the garden, there is something tremendously reassuring about being in the persistent presence of a plant with such an irrepressible urge to grow, claiming ground at a rate of 10 metres (32 ft) a year. A single plant of *Rubus fruticosus* or one of its many subspecies would, if planted in the centre of a soccer pitch, have reached out to embrace both goals by the end of its second growing season, stems arching and plunging across the ground, rooting where they touch the soil in a process known as layering. But the brambles in our gardens are not just hovering about on the edges having lolloped over the fence from the neighbours' side – they're popping up in every bed, for which contribution to the gardener's workload you can probably thank the birds and the occasional furry critter. Layering can only spread a bramble so far; to enlist the help of creatures over a wider territory, this prickly interloper relies upon the deliciousness of its seed-laden fruits – those dark purple berries for which, for a few weeks in the whole year, we would bless the presence of the brambles in our garden, at least until Michaelmas Day when the devil is rumoured to pee on them (Satanic micturition aside, it's certainly the case that, even if you can find a blackberry beyond the first week in October that's not covered in fluffy grey mould, it certainly won't taste very good). Most brambles being polyploid – that is, having more than the two sets of chromosomes you or I walk about with – their opportunities for genetic variation are greatly increased. And yet the associated condition of apomixis also allows them to produce viable seed without the need for fertilization should environmental conditions dictate. Able to flip between sexual, asexual, and vegetative methods of reproduction, it's no wonder that quite so much of my time in the garden is spent in the company of these adaptable plants, their first year, non-fruiting primocanes prodding and probing at me with the greatest urgency as I go about my business, the second year floricanes having better things to do with their time, such as producing flowers and fruit, though still more than capable of tripping an unwary gardener in passing. In wild and semi-wild, managed environments, brambles perform a whole suite of functions; protecting sapling trees from overbrowsing by deer, providing nesting habitat for the wren, thrush, and nightingale, and food for over 200 species of insect, as well as the deer, dormouse, fox, and badger. In the garden, these brambles (eventually) give me jewels to brighten summer puddings or crumbles, something at once tart and sweet to have with my breakfast, (rather more regularly) leaves for my tea and, by the end of the day, a collection of grazes, scratches, and splinters. I wouldn't have it any other way, directing towards them alternating waves of awe and fury in the knowledge that, at the first sign of my relinquishing hold upon the garden, nature will send the brambles rolling in.

The word "briar", loosely wielded, can encompass within its precincts not just the bramble, but all manner of unruly scramblers whose thorny stems enjoy the company of others of their kind, forming, if left to their own devices impenetrable thickets, as well as the briar of old smoking pipes, carved from the hard and heat-resistant roots of the Mediterranean tree heath *Erica arborea*. It can be applied with equal justification to the Mexican sarsaparilla vine or "greenbrier", *Smilax aristolochiifolia* – the word "sarsaparilla" being a derivation of the Spanish name for the same plant, combining the words *zarza* (bramble) and *parrilla* (little grape vine). But most definitions of the word briar include a specific reference to the wild rose and, in the context of a winter spent wrangling spiny things at the edges of gardens in the south east corner of the UK, that's the flower I hold in my mind's eye as a promise of sweetness and grace to come, pleasantly entertaining the anticipation in order to distract myself from the ferocity of heavily thorned stems snatching at my hat, whipping against my face, and snagging in the skin of my forearms. Such reveries would seem to have been keeping company with roses for centuries, the King of the Fairies in Shakespeare's *A Midsummer Night's Dream* identifying two wild species in his depiction of Titania's dreamy bower, where Oberon's appropriately flowery description embraces both "sweet musk roses" *Rosa moschata,* and "eglantine" *Rosa rubiginosa* or "sweetbriar". The one white in bloom, the other a pale pink, to accurately imagine a wild rose in flower having never previously seen one would be an achievement. There's none of that combination of demureness wrapped about in ballgown layers that we've come to expect from a florist's rose, no elegant goblet form, no ruffles. Just simple, five-petalled flowers, gently dished, a yellow disk in the centre surrounded by a forest of splayed stamens, leaving nothing to the imagination. Demure, the wild rose is not.

There are no flowers at this time of year, just stem and thorn, and what the birds have left of the fruit, the bright red hips with their jammy innards, hedgerow superfood, packed full of antioxidants and fibre, a rich source of Vitamin C. A mature dog rose, *Rosa canina*, is woodier by far than its distant relative, the bramble, with a framework at once more structural and longer lasting. A bramble climbs into and through a hedge, tree, or thicket, and even when it binds together, it retains something of the air of an interloper, whereas the dog rose – due to the framework of its enduring, perennial stems – feels like part of the furniture. And while we might, with Shakespeare's Juliet, believe "that which we call a rose by any other name would smell as sweet"[7], there's something about the "dog" that intrigues and, whether it's there as a relic of this briar's reputation, first suggested in Pliny the Elder's *Natural History* (77 CE) as a cure for the bite of a rabid canine, or the resemblance of its thorns to the curved teeth of that animal, I've always felt there to be something dismissive about the adjective as applied to a wild and unruly plant, the poor relation to the gorgeous, cultivated things of the formal garden.

We know that we can chop and hack away at this brutish thing with impunity, confident in the knowledge that it will bloom and fruit again next year, but when it comes to the care of its more rarefied cousins, uncertainty takes a hold. Mystery has grown up around the art of rose care to the extent that the new gardener often finds themselves paralysed with indecision over what to do with them. I have always found roses to be marvellously tolerant of my sometimes-inexpert ministrations and have never been tempted to waft around the borders in summer, a hostage to intricate and restrictive spraying regimes, as traditionally seems to have been the recommended course. For the truth of the matter is, notwithstanding drenching the things in noxious chemicals, the fungal black spot that disfigures the leaves is an inevitability in all but the most resistant of cultivars. Since the spores of the fungus responsible persist upon dead foliage and can be spread through the splashing of a raindrop, the sensible course of action with roses is to ignore the imprecations of a previous chapter and practise scrupulous hygiene when it comes to the collecting up and disposing of spent leaves, which should never find their way into any compost heap but meet their end in the heart of the bonfire.

But while black spot might be unsightly, it won't kill a rose any more than will inept pruning though, trained in the traditional manner, I will probably always look to place my pruning cuts just above an outward facing bud, and persist in removing dead and spindly growth, even though there's a strong groundswell of opinion that the roses don't really appreciate such niceties, and wouldn't give a fig if you were to come at them with a hedge trimmer (see *A cut above*, page 170). But hedge trimmers are for hedges and, in the borders at least, I'll continue to prune with secateurs. Those moments spent in contemplation of a rose's form, holding each stem, and carefully determining where to cut, seem to be far more conducive to a meaningful relationship than would be the case were one party to be applying a pair of noisy, reciprocating blades to the other. A more careful approach avoids the chance of administering a "short back and sides" to your rose collection or, worse still, a buzz cut – either of which actions would deny something of the rose's nature as a plant particularly adept in growing long and graceful stems, while simultaneously relinquishing one of the great pleasures of gardening in winter, that of persuading its growth to conform to your own notion of order, a mindful occupation which can be accomplished in as ornamental a fashion as you have the time and inclination to achieve. For the first year's growth on a rose is a malleable thing and, as long as you leave yourself sufficient material to manipulate, there's an opportunity to create the kind of visual interest and structure that will help the plant achieve its fullest potential, even despite itself. Since plants tend to take a direct route towards the light and flower at their apex, it is the gardener's job to introduce diversions wherever possible, encouraging the rose's stems towards the horizontal or bending them right back around and down, the better to induce the production of flowering shoots. Tradition would have you tie your rose to a support of some kind, a trellis, obelisk, or wires upon a wall but, as long as the twine is sufficiently tight to prevent chafing of one stem upon the other in the wind (and always twine, never metal or plastic ties that will act as a garrotte upon the stem and choke the life out of it), there's nothing to prevent the rose from being tied to itself, repeatedly, creating forms of increasing complexity, sculpture in living wood for the winter, wreathed in flowers by summer.

And where once such an extravagant, flowering framework might only have been seen adorning a wall or wrapped around a more traditional support, increasingly the display is being persuaded back down towards the ground within the mixed border, jostling with phlox and delphiniums as a counterpoint to the weighty, evergreen balls of box and of yew, a welcome antidote to the bare-legged sterility of the traditional rose garden.

There's an aspect to the garden that many of us find challenging to create for ourselves, due in no small part to its being a tricky thing to articulate. But having teased it out in conversation with so many people over the course of my work, I've come to recognize a desire for something that could be described as a certain *presence* to the planting – attainable through the inclusion of elements that give feelings of height, of volume and of mass; properties which, in less manicured settings, ivy, bramble, and briar need no help to achieve. But most of us want something a little less threatening than a temporarily tamed thicket giving body to our gardening ambitions – to conjure that feeling of enclosure, safety, and permanence without the teeth and thorns. Though we'll accept, for three months of the year, the elegant airy cages of the domesticated briar, as long as it's quick to cover itself in foliage and flower as the days begin to lengthen. If we're also demanding solidity from those specimens that attend upon us reliably from one month to the next, we're going to have to augment those artful filigree constructions, fill in the gaps and look towards establishing an enduring structure with the dense and evergreen.

How to DIG A HOLE

You don't want a hole. Not really. No sooner have you dug it, you'll be putting something in it – a plant, a post, or a pond. And then, there'll be no hole. This is how it should be – holes being by and large not to be desired, which is why the universe takes such pains to fill them up the moment they appear. The Undesirability of Holes is a good place to begin, before you set about the business of digging a new one.

Nature is altogether more subtle when it comes to planting, tending to "introduce" rather than "insert". But here we are with a plant (or a post, or a pond) and a wish for it to become a part of the garden, so a hole must be dug.

C*onsider the ground where you would like your hole to be.*

Put the spade aside and take a step back, the better to survey the non-holey wholeness of the soil into which you're about to sink it. Take a moment or two to appreciate the lack of interruption distinguishing its integrity. Into that unity, you must now make your intrusion.

Begin with a pause, and a deep breath.

Take up your spade and place the blade firmly where you would dig, a foot on the tread, your leading hand on the handle, the other on the shaft.

Now, resisting the temptation to let apology cloud decisiveness (for an apologetic hole is of no use to anyone), drive firmly not from the knee, nor even from your hip or your elbow, but somewhere a foot or two above your head and slightly to the rear of your shoulder, until both feet are firmly on the ground.

Pull gently backwards, scoop and lift, then turn and drop – the more fluidity and grace you can bring to this solemn and soily dance, the more your bones will like you in the morning.

Repeat, until a hole of appropriate size appears.

Some people will tell you that digging a hole is all about the right spade and thick-soled boots. About keeping a straight back, bending your knees, and engaging your core, and about the theory of levers. It is about these things. But not all about them. Not even mostly.

It's about making interruptions and taking liberties, sticking a dirty great blade deep into a nurturing friend. And so, it's really about asking pardon and permission. And the understanding that though we might hope for both forgiveness for our actions and blessings upon our labour, we have no right to expect them.

Ever green

We all know those characters who can be great company for an evening – the larger-than-life types, the fabulous raconteurs, the beautiful ones. Surrounding ourselves with those who would dazzle our senses might seem an attractive proposition for a short while, but eventually the glitz and glamour fades, morning comes and all that remains is a house full of hungover guests rocking eye-bags and last night's makeup, and a whole lot of tidying up to do. These are the moments we begin to appreciate the understated magic possessed by those solid, sensible, and steady fixtures in our life, appearing fresh-faced and unflustered while bearing strong coffee, pastries, and bin bags. They are our evergreen friends, and they have their analogue in our gardens.

Persistent, reliable, ever-present in the background, they demand little, remaining unremarkable for most of the year and stepping forward only when needed, like now, in winter, when all else in the garden is either sticks or mush. Your evergreen cover has got your back, the low-maintenance building blocks of the garden, the plants we use to define and to partition the space. It's evergreen hedging we turn to in embracing the concept of "garden rooms" – distinct zones with different moods, personalities, and functions – for its reassuring bulk, solidity, and steadfastness. Structure in the gardens of the great and the good is comprised almost entirely from either of just two plants – yew, or box – building blocks for towering, buttressed walls, looming columns topped with fantastical figures, disciplined edging, and the intricate weave of the knot garden. The lessons we learn from their employment within such impressive settings can be refined and reduced to make sense in our own back gardens and even, in the case of topiary, on our balconies.

For lower hedges, those supremely well-manicured emerald ribbons edging beds or accompanying the visitor as they travel along the garden path, the shrub of choice has for centuries been box (more commonly boxwood in the US); typically English box (*Buxus sempervirens* 'Suffruticosa') or, with its even more tightly packed and diminutive foliage, small-leaved box (*Buxus microphylla*), though a wider range of varieties are drawn upon in climates that swing between burning summer heat and winter frosts hard enough to recommend the swaddling of hedges in burlap. On chalky Mediterranean soils, box exists in its natural state as a shrub of the woodland understory, in poise more relaxed than might be recognized by the garden visitor accustomed to meeting it within the context of those well-manicured spaces with which it has become affiliated. But over time, regular clipping can be relied upon to increase density to the extent that, on the occasions where I've found myself absent-mindedly placing a mug down to rest on a newer section of hedge, I've had the upsetting prospect of watching my eagerly anticipated refreshment disappear through the greenery. Had I chosen as a coaster more mature, more frequently trimmed growth, the mug would have felt itself better supported, and the gardener not so foolish. It's this very property of box, its small leaves set naturally close and becoming more so with every cut, that has made it such a versatile, biddable material with which to work, ideal for shaping into topiary or hedging, populating the garden with living, green furniture, at once wall, boundary, and ornament. But in recent years, box has experienced a challenge to its reputation as the go-to plant for low-maintenance hedging and topiary, an assault on all sides from rust, blight, mussel scale, and moth. And, while sections affected with the characteristic orange-brown pustules of the rust fungus can be cut out, and mussel scale insects treated with the same kind of diluted soap or neem-oil based spray we might use to control scale insects on houseplants, the measures required to repair and protect against damage caused by caterpillars and blight, while not heralding the end of box in gardens, have lifted it into a higher maintenance category. Even where the care regime is limited to organic pesticides – with the addition of a little copper mixed in with a nourishing dose of seaweed-based plant tonic – it's questionable whether many of us will want to commit to spraying plants regularly through the growing season. Though here we should pause and consider that, for all the years we have been clipping our hedges into submission – whether box, yew, or the privet of the garden suburbs – how much time have we spent feeding them? Having bits of your living self regularly removed is a stressful experience, and the best horticultural instruction insists that all pruning activity (including clipping) should be followed by the courtesy of presenting the subject with a decent meal. We have taken our box for granted. I can't help but wonder if we should have been expecting it to show signs of vulnerability, and perhaps the only surprise is that it held out for so long. And so, while I hesitate to plant new box hedging just now without giving careful consideration to its ongoing care, I feel obliged to nurse any ailing specimen I come across back to health.

Box is just one among the entire panoply of evergreen plants available to the gardener, and while the horticultural trade has yet to find an issue-free default substitute, there are plenty of contenders including Wilson's honeysuckle (*Lonicera nitida*), many varieties of pittosporum, Japanese holly (*Ilex crenata*), and the barberry *Berberis darwinii* 'Compacta'. But it's yew, for so long box's companion in the more formal areas of the garden, that most reliably makes the grade.

No tree was born to be a hedge, a cone, a sphere, a monolith with a peacock or a squirrel sitting atop, or any other of the many indignities to which we subject the evergreen plants in our gardens, but the yew (*Taxus baccata*) seems to take it all in good spirit, which is all the more remarkable given its pedigree. For the yew is the ancient tree of churchyards, more ancient even than the churches whose entrances it so often adorns, frequently erected upon the meeting places of pre-Christian societies where the yew tree was held in high regard as a symbol of longevity, death, and rebirth, and in whose presence the faithful would gather to worship. A handsome tree, its cinnamon-coloured bark growing gnarled and fissured with increasing age, with deep green needles and bright red, berry-like fruits, it is one of only three coniferous trees endemic to the UK (the others being Scots pine and juniper). Every part of the yew – excepting the fleshy red aril around the seed – is highly poisonous, the organic compounds (or taxanes) responsible a key constituent of widely-used chemotherapy treatments. Such toxicity often causes alarm when yew is suggested as a plant for the domestic sphere, though no one seems to give a thought to the serious effects that munching on a salad of snowdrops or daffodils could have on them, and the chances of mistaking a yew tree for a spring onion are far lower by comparison.

Conifers are not renowned for their forbearance when it comes to hard pruning, with most refusing to sprout anew from old wood. It's a testament to its usefulness as a garden plant that the yew tree will not only suffer itself to be planted rank and file as a hedge, then clipped into all manner of tightly defined and outlandish shapes, but will also respond to being cut back with the production of fresh, needle-clad growth. That it seems content to supply the backdrop to our floral displays, only coming to the fore in winter when the garden is bare, is a further reason why I find it hard to imagine gardening without this most versatile of understated plants (it might not be happy to have its feet standing over-long in water but, since we share the aversion, I'm inclined to forgive).

From a distance, the way light plays over the coniferous yew is quite different to the effect seen with glossy, broadleaved evergreens. Whereas the healthy foliage of box plays with each photon striking its surface, sparkling in the sun and bouncing rays around the garden, yew appears to capture the light and envelop it deep within itself, holding on more protectively than jealously, a deeply green benevolent presence at the back of the border. You can't help but feel that the heart of a yew is a dark but comforting place, one into which I've yet to lose my tea, though that's probably less to do with any perspicacity on my part than it is with the yew's lending itself to hedging on a far grander scale than does box and presenting fewer invitations to carelessness.

For most of us, the days of wriggling our way into the middle of hedges are long behind, unless it's to get to grips with an errant bramble, and it's the face that the yew presents to the world with which we're most concerned. The scope of garden architecture possible with these verdant elevations is breathtaking, whether maintained in angular fashion with straight lines and crisp corners or sculpted into more organic shapes. It can be seen in the 300-year-old hedge that rises from Cirencester Park in the Cotswolds like a huge green soufflé above the town, an immense semi-circle reaching 12 metres (40 ft) high and over 45 metres (150 ft) in length and taking two gardeners a fortnight to trim each year. At Doddington Place in Kent, the huge cloud-pruned yew began life as a formally clipped feature until, having slipped into disrepair during the Second World War, the decision was made to celebrate and maintain its billowing curves. Every honest gardener will be forced to admit that at least some degree of the appreciation experienced when facing the horticulturally spectacular rests in the knowledge that its ongoing maintenance is somebody else's problem. Thankfully, a yew feature in your back garden or in mine is likely to require little more from us than a gentle trim two or three times a year, depending on whether we like our outlines crisp or wafty, though in this case at least, we can have our metaphorical cake and eat it, crispness naturally transitioning over time to waftiness, with a trim being administered before the word "shaggy" could be used with any justification.

Partly, it's the ubiquity of box and yew that justifies spending so much time in consideration of their virtues and drawbacks. Partly, it's that the principles governing their use and ongoing care are central to a less intensive approach to the garden, acknowledging that while there is really no such thing as low-maintenance gardening, surely, if you want to garden by doing next to nothing, you will recruit some evergreens to the cause. To this end, it's worth remembering that not all evergreen plants are equal when it comes to the burden they place upon the gardener; that the profusion of colourful berries on the firethorn or evergreen varieties of cotoneaster, for example, needs to be set against their habit of year-round foliage renewal and the attendant necessity of having constantly to tidy up after them. But, thankfully, the list of evergreen plants exhibiting a more steadfast attitude to their leaves is mercifully long, though not all exhibit the staying power of the Great Basin bristlecone pine (*Pinus longaeva*) on the west coast of the US, maintaining a grip on its needles for up to 45 years. Presently wafting its gorgeous scent around the garden, quite what the Christmas box (*Sarcococca confusa*) does with its leaves is a mystery to me; I've never seen one anywhere but on the plant or sometimes, when cut for a vase, indoors upon the mantelpiece, and it's been here for 15 years. Sarcococca, left to its own devices, might not possess the most disciplined of habits, being somewhat inclined to wave its arms about despite a pleasingly fulsome stature, but the prospect of sunlight gently sparkling off its handsome two-tone foliage – deep and glossy above, a paler satin below – enchants the eye as much as the scent tickles the nostrils. Clipped, it will never quite attain the precision formality of box, but neither will it suffer from blight or moth.

Quite how much time to spend clipping plants is a matter each gardener must decide for themselves, modifying their plant choice and garden aesthetic accordingly. Undoubtedly, a desire for control is involved, and there's catharsis in being able to impose an order, however temporarily, that seems wanting in other areas of life. Some may develop a nascent interest in topiary or cloud pruning, seeing hidden in every shrub, forms abstract or geometric that need to be released by the editing out of obfuscating detail. For others, trimming a hedge once or twice a year might be sufficient to restore equilibrium to both mind and garden, while neither fault nor failing could be found with those who manage to resist the impulse to clip anything at all from one month to the next, finding delight in the natural forms and habits of growth of every plant in the garden. The temptation to consider this versatile category as so much "outdoor furniture" can be hard to resist but, useful as structural elements with year-round presence, there's so much more to evergreen plants than a narrow consideration of hedges and topiary forms would suggest. Still with half an eye on the functional, but keen to delight the senses too, we can choose flowering evergreen climbers to clothe walls, screen less-lovely aspects of the garden, and play hide and seek with the best views as we move around the garden, while also charming us with scented blooms throughout spring (*Clematis armandii*) and into summer (*Trachelospermum jasminoides*).

If we can allow ourselves simply to enjoy the presence of these plants for their own sake, rather than appreciating them merely as raw materials compelled to perform a particular role, we will discover a rich, albeit monochromatic world of shape, form, and texture that can add depth to our gardening experience while asking little more from us than our occasional attention. Conditioned as we are to crave the titillation of flowers, this might pose a challenge, but even here evergreens have made provision, from the subtle but gorgeously scented blooms of sarcococca or daphne in the winter garden, to summer's huge waxy flowers on the southern magnolia (*Magnolia grandiflora*) and, if such understatement doesn't scratch the floral itch, spring brings the showy display of acid-loving shrubs such as the rhododendron, in shades of every colour from reds and oranges, through yellows to mauve. Camellias too, though for a fortnight or so each year, a white-flowered camellia becomes one of the neediest plants in the garden, requiring constant deadheading in early spring if there's so much as a hint of frost, at which point the sophisticated shrub with its white blooms will resemble nothing so much as an old bush from whose branches someone has painstakingly hung a collection of used tea bags. Red- or pink-flowered varieties, however, remain largely trouble free throughout the year, providing they are kept from drying out in summer, or getting a sniff of the chalky, alkaline soil, on which any attempt to grow them will see their leaves turn yellow in disgust.

Those evergreens that don't pop quite so gaudily are among the most soothing of plants in the garden, schemes that focus largely on foliage rather than flowers possessing a capacity to invoke within us sensations of restfulness and calm. Whether we attribute this to an awareness of the modest calls they're likely to place on our time and energies, or to the widely accepted reputation of the colour green for influencing our mood to the better, the inclusion of something more than a token evergreen presence in our gardens seems to make a good deal of sense. And so, intent now upon company and captivation rather than control and coercion, we can put away the shears, finding ourselves free to choose plants whose mature size is more in keeping with the scale of our gardens, our selection criteria depending less on how well a plant is likely to respond to being clipped into a cube or a cone or a ball, but on how well its natural form will look in this spot or that. We might choose from any number of handsome and reliable specimens – the character actors, full of personality and attitude – though beauty is very much in the eye of the beholder with many of the plants that find themselves in this category. You might admire the mahonia for its indestructible nature, its upright stance, the holly-like leaflets, or its long sprays of tiny yellow flowers borne through the winter and so greatly appreciated by insects resisting the call to hibernate but taken together, these properties have made it an ideal candidate for landscape planting en masse. A snobbery grows around plants that have become too familiar, and mahonia (*Mahonia* x *media*), along with its close relative the spiny-leaved barberry (with whom it shares many characteristics, not least the surprising saffron yellow colour of its heartwood when cut), the broad-leaved *Photinia* x *fraseri* 'Red Robin' with the ruby flush of its new foliage, the spotted laurel (*Aucuba japonica* 'Crotonifolia'), cherry laurel (*Prunus laurocerasus*), Portuguese laurel (*Prunus lusitanica*) and the many varieties of Japanese spindle (*Euonymus japonicus*), are all frequently derided as "car park plants". And, while you will doubtless find many a specimen in such locations or planted far too close together with others of its kind by the walls of newly built apartment blocks, to dismiss them purely on the grounds of their ubiquity would be rash (I've yet to see a garden at one of the major flower shows that lists on the design brief a requirement to use only car park plants, grown to their most handsome and undeniably ornamental effect, but I live in hope). There will always be choicer more rarified specimens with which to dazzle visitors to your garden, but exclusivity has its downside. Those plants that the gardener finds no difficulty in getting hold of tend to be the ones that are easy to care for, with the converse often being true for the less readily available and so, when looking to populate our borders with evergreen character, we could do far worse than taking the commonplace as a starting point.

Fads and fashion also play a role in the garden, and nowhere more so than with our relationship with conifers, the majority of which retain their needles or scale-like leaves throughout the year. This modified foliage is thought to be an adaptation to cold or dry climates and retains water more efficiently than the broad leaves of more recently evolved deciduous trees, while being better at shedding snow and permitting the passage of the kind of high wind that might otherwise fell a tree. The conifers' symmetry and scale, the scent of their resinous sap on a warm day, the sheer variety of form in flower and foliage and the cones that house the naked seeds all suggest an ideal candidate for inclusion somewhere in the garden, given sufficient space for such majesty to be appreciated. And yet, perhaps in a parody of the activities of the Forestry Commission whose plantations of Sitka spruce and Corsican pine at that time still marched in rank across the countryside, it was hard to pass a garden in the 1970s without seeing a motley assemblage of dwarf conifers, planted on or in close proximity to a rockery – both trends that have witnessed their zenith come and go (and which may therefore be entirely ripe for imminent revival). Around the same time, horticultural trade did a fantastic job of selling the benefits of the Leyland cypress to the general public as an exceptionally fast growing and not altogether unattractive tree, entirely suitable for hedging, though the messaging regarding their appropriate aftercare was communicated with less clarity. With a growth rate of over a metre (3 ft) in a single year and a mature height of more than 15 metres (50 ft), a hedge of *Cupressus* x *leylandii* requires cutting at least once, preferably twice a year to keep it in good order and, unlike yew, strongly objects to being cut back beyond the last few years' growth, leaving unsightly brown sections where pruned with excessive zeal. Consequently, though still widely planted, the "leylandii" has become the cause of no little neighbourhood acrimony, and the subject of countless boundary disputes lodged with local authorities.

Forty years on, the leylandii hedge my old Da planted around the front garden of our house is still there and, though the neighbour seems to prefer his side shaggier than the current resident, it's not been allowed to extend beyond the modest dimensions it achieved within the space of its first year. For a North London lad whose visits to see the festive lights in Oxford Street would customarily end happily munching on hot, roasted chestnuts and gazing up at a huge and handsomely fragrant Norwegian spruce in the centre of Trafalgar Square, the only surprise about that other conifer-related blunder is that people wouldn't think carefully before planting a similar specimen in their back garden. But a Christmas tree plunged into the ground – roots and all – to save it from the dustmen, the bonfire, or the shredder can in time quite easily result in a thing 8 metres (25 ft) wide and 9 metres (30 ft) in height, prompting anyone within toppling distance to regard it askance whenever the wind gets up. Undoubtedly the process of giving the tree an annual root prune, digging it up and dragging it indoors for those few winter weeks, would keep a check on such alarming vigour, but it's hard not to wonder if the Silent Generation responsible for the gardens of my childhood were simply exhibiting typical caution in their clear preference for nothing more troublesome than the quietude of rockeries, and the gentle soft, green composure in an agreeable collection of dwarf conifers.

We began the chapter likening the evergreen in our garden to a reliable friend playing a decidedly supporting role, allowed ourselves to consider them for the part of character actor – a little humour and humanity injected into proceedings aside from the flow of the main drama – and found ourselves in a place where, given the opportunity, they could sparkle and shine in their own right, rising above all others as stars of the show. But a garden, as Gertrude Jekyll would have us know, is a grand teacher[8], and to leave our consideration of evergreen plants at the point of knowing how we can use them, without stopping to think what we can *learn from* them would be to ignore, almost wilfully, the greatest lessons they would teach, through careful observation and emulation of their finest qualities. Within the restrained majesty of the yew that suffers itself to be moulded into forms to suit the gardener there is modesty; self-assuredness in the handsome foliage and proud stance of tall shrubs content to watch other, more ephemeral things take the limelight until a time of year when almost nobody is around to witness their own hour of glory. There's dogged insistence in the importunity of the sarcococca that invites us, with the gorgeousness of its perfume, to step once more into the garden after so many weeks cooped up indoors. Ultimately, that which is ever green shows up day after day, careless of season and weather, relentless in the pursuit of its objective, thrilling with a force that just won't quit until the fall of the final curtain, and giving us the inspiration that we need through the dark, cold days of winter to see us through to spring. The fundamental lesson of the evergreen is one that binds up strength and humility with resilience and determination, and that is a teacher in whose presence I am content to stand and stare.

SHOOTS

to the gardener

I have slept, and sensed your gentle presence throughout wakeful winter dreams, but now the time is come to rise, to stretch, to grow, to greet the bright new year with promises fulfilled of leaf and bud and bloom. And, oh! The joy of *life*, and spring and brightest green! The chance to flex and fill, extend, unfurl… The Taking Up of Space.

The waiting's done.

Will you come along again? All these, though things that I can do alone, so much more *fun* with two.

New beginnings

Without doubt the most keenly anticipated of seasons, spring is the time when many of us remember our gardens. Perhaps we've only ventured through the back door these past few months to fetch an armful of logs for the fire, but inevitably the day comes when we pause outside, senses quivering, sleepy synapses firing faster now, alert to change in the air. Lured deeper into the outdoors by the promise of warmer weather and brighter days, we cast about for signs to assure ourselves of winter's passing, attention drawn by the flowers of hellebores and snowdrops we forgot we'd planted last year. In truth, they've displayed both bloom and a distinct indifference to the prospect of our grateful notice since the middle of winter, and quite which season can properly claim them as their own is as difficult to discern as the moment at which one hands over to the next. But by the time crocuses and daffodils stage their cheery display, we feel confident in proclaiming the arrival of spring, at which point, likely as not, it promptly snows. Nature does not like to think of herself as predictable, and our gardening partner has a sense of humour.

Desperate to feel the warmth of the sun, we become highly attuned to changes in the weather and, regardless of the chill breeze, that first weekend when its rays seem to transmit not just the watery light of winter but the power to raise the temperature of our bodies by a fraction of a degree, we're out of the house, often to be found standing in bemusement amidst our winter wreckage. It is a day of reckoning for our diligent neglect, this tangled hotchpotch, product of our wildlife-friendly endeavours that rendered service as restaurant and boarding house to a thousand unknown selves, but whose usefulness recedes as the need for emergency winter rations and shelter diminishes by the hour. Time to feed the compost heap, to fuss, primp, cut down and rake out the old to make room for the new. The days draw out, the soil warms, the sap rises and there's a window in which, as the garden bursts into life all around, you can indulge in the feeling of being in control, of knowing what you're doing, of being in a state of flow. It's an illusion – summer will set you straight – but right now nature feels biddable, every plant, every border, every view full of potential, and you at the centre, directing it all into being with thrust of trowel or a wave of secateurs.

In reality, the moment when each season fully inhabits itself is fleeting, and most of our time in the garden is spent upon the threshold as one transitions to the next. Winter, being loath to leave, maintains a stubborn grip upon proceedings long after the outriders of spring have been welcomed, and any notion of life in a liminal zone of hazy, dreamlike ambiguity can be forgotten. While the veil between one world and the next might be vanishingly thin, here on the boundary between winter and spring a muscular contest is fought out, one where we might be forgiven for wondering if the chilly grasp of the old will ever weaken. Here, in this mainly mild and temperate place, the emerging shoots of bulbous perennials pushing their way up through the soil and into the light will know that what awaits them could be rain, snow, or sunshine, and quite likely a mixture of all three. At the same time the gardener, accustomed to weeks of buffeting wind alternating with short spells of unexpectedly warm sunshine, embraces a stratified approach to outdoor attire, progressively peeling off waterproofs, woolly jumpers, and microfleece over the course of the day and ready to scrabble back into each protective layer according to the demands of elemental caprice. While the vagaries of the British weather recommend a degree of flexibility and preparedness in clothing, this peculiarly horticultural variation on striptease is so closely identifiable with this in-between time at the beginning of the year that I realize I have, almost by default, begun to think of the interregnum as a season in itself – *winterspring* (though "sprinter" has been suggested, what that portmanteau gains in catchiness it loses in chronology) – and that with this minor act of reclassification, an unexpected tranquility has stolen in upon me. Impatient neither to shuffle off the previous year nor run headlong into the next, a space opens out around me and there seems time to appreciate the here and now. Where changeability itself is a constant, and a reassurance to be both relied upon and settled into, there's as much comfort to be taken in such familiarity as there is in surrendering – at long last – to the impulse to cut down and tidy; to be for an afternoon an agent of immediate transformation. With a contented sigh, I take up my secateurs and get to work.

But when it comes to the whims of the weather, the process being played out in nature reflects our own inner meteorology as we grapple not only with steering a course through a crowded field of possibility and potential, but with deciding what we take forward into the new growing season, and what we leave behind with the last. Because, for all the January fuss, the parties and the resolution making – and notwithstanding the importance of winter to the plants in the garden – the world has seemed in limbo since then, and still not feeling entirely free of past entanglements we can be forgiven for finding it hard to gain traction with new plans. Occupying a unique role in our home life, our gardens become imbued with meaning; over time, we develop a talent for taking portions of ourselves – our hopes, dreams, even our joys and sadness – and investing them into individual plants, objects, scenes, or schemes, whether real or imagined. The price of such emotional projection is that the success or failure of any horticultural endeavour can often be felt more deeply than might be understood from a purely objective consideration, and this lies at the heart of our ability to make progress with all our gardening plans. A certain ruthlessness may be called for when reviewing the feasibility of our intentions.

Only recently struck by a realization that, much as I love the magic of growing from seed, I'm temperamentally far better suited to raising hardy perennial plants from cuttings than tending to trays full of delightful but demanding annuals, I'm now resolved to exercise self-control in the vicinity of seed packets and catalogues, while all about me are loudly trumpeting their germinating ambitions. And so, we bring a level of detachment to our assessment, not only of our own strengths and weaknesses, but of what plants we intend to grow, which beds we mean to extend, what structures or elements of design and layout we hope to modify – and what better time than now to measure each of these goals against our available resources of time, energy, and money? As the days lengthen, we have benefited from the clarity and space of winter in which to wrestle with the thorny issues of what to cull from our gardening lives, any frantic scrabbling and threshing against our consciences representing nothing more than the last desperate throes of the rejected. Separation can be painful, but ultimately, necessary if we are to grow, and flourish. It ends here, and we move forward into the new year which, by all that's sensible, should really start now, in spring, rather than in mid-winter.

*

But step into the garden with me again and stand here till the traffic's faraway thrum fades in our ears to a murmur, softens to a gentle drone. All around bright chatter from the throats of tiny birds, distant honking of geese on fields and overhead the gulls, beyond the river's tidal reach and quite at home so far from sea or spray as droplets of a different kind betoken rain, though still a way away in any force and, felt now on the skin not rain, but wind, in faintly perfumed gusts. Sweet soil, now sarcococca, still; now… coffee? Coffee's bitter note suggests the lime-green exclamation marks of Wulfen spurge are mustering their energies for display, chartreuse over bluest green, with tulips and forget-me-nots as understory players. But not quite yet. Here, now, fat carmine paeony buds protrude, soil rudely thrust aside and forgiving or oblivious to discourtesy, settled into place once more; and everywhere the weedlings, speedwell and cleaver, stealing a march on their weedy friends, taking full advantages of open ground. We came outside to smell the freshness of the earth and hear the birds but now, we realize the scene is set, and guests assembled for something of more significance, and we're the main attraction. And suddenly we know (as if we've known it all along) the time is right to recommit ourselves to this space, this work, this *life* that reflects such meaning into our own. And so, we bow our heads, renew our vows, and pledge to grow again.

*

Reluctant as our seasons seem to conform to any timetable, the meteorological calendar (the irony won't, I'm sure, be lost on anyone) seems particularly poor at forecasting their passing. Perhaps it's the contrarian in me, the British habit of rooting for the underdog, or the appeal of the year being neatly partitioned into perfect quarters, but I persist in following this admittedly terrible system and calling each season in at the beginning of the first of the month (March, for spring, then at three-month intervals), invariably hopelessly previous. I'd be far better waiting three weeks for a solstice or an equinox and, if not absolutely guaranteeing, at least standing a fair chance of the season's announcement coinciding with some characteristic weather patterns and floral action.

But eventually the frequency of wintery interludes diminishes and, surrounded by the burgeoning signs of new life, it would take a conscious effort to resist the positive energies that seem to radiate from every tree, rock, and clod of earth with the onset of spring; that lightening of the heart that follows with inevitability, even in our darkest times, the knowledge that the year has finally turned towards the light. In part, it's relief after the lengthy darkness of winter, in part it's the excitement of meeting old friends we've not seen for a whole year. And part of it is undoubtedly gratitude for a chance to begin again. This continual reset and start over, a wiping clean of last year's horticultural slate, taking forward from our failures only lessons, and rolling over to the new account all our previous successes, flows throughout the relationship we have with our gardens, bringing with it turgor and nourishment as the sap now pumping up from the roots of every plant animates each stem and branch and bud. Is it only in your imagination that you can hear this enlivening substance, an *aqua vitae* more quintessential than any spirit to which the name was historically applied, coursing through a million tiny vessels all around, or is it only with a stethoscope or an adapted microphone held against the trunk of a tree that you can hear it gurgling away within? For ears more sensitive than ours, the spring air must be filled with a cacophony of new life; below the busyness of birdsong, an incessant percussion of bud scales being forced apart in the eagerness of their contents to greet the new day, the constant squeaking of fresh green growth as stem and leaf creak noisily past one another to assume their proper positions.

And with the arrival of the season in all its fullness, a new confidence is discernable in the garden, as it is in the fields, hedgerows, woodlands, and every scrape of soil within the urban landscape; not so much a determined drive, which would be suggestive of effort, as that kind of settling back into a long forgotten state of flow – such as might be experienced after a period of injury, or enforced rest – where simply to *be* is to experience the delectation of the power to create. This momentum, building winterlong and finally finding its outlet as daylight lengthens and temperatures rise, emerges in a joyful rush, earth laughing in flowers but with none of the knowing judgement of the American poet and philosopher Ralph Waldo Emerson's "Hamatreya", an innocent delight at the rediscovery of such awe-inspiring power to call life out of the ground. The breathtaking speed at which bare soil becomes shrouded in green, naked branches are clothed with leaves, and lawns begin to clamour either for attention, or for licence to grow into something altogether more exciting, all serve as a reminder – should one be needed – that we've reached a milestone in our journey through the year, one of the points in the gardening calendar at which the pace of change in the beds and borders undergoes a significant shift. There will not be another until midsummer when the growth rate starts to ease and, until then, we are surrounded by evidence of nature's single-minded purpose – to grow – presenting us with the challenge of how to react. Because, notwithstanding the studied indifference with which we have come to treat the advent of spring, having begun to engage once more with our natural environment our first instinct, and a not unreasonable one at that, might be to stand a while and watch it all take place around us; to pause on April woodland walks and marvel at the violet bluebell haze, the misty layer hovering calf-height above the ground, returning to our gardens to find forget-me-nots have clubbed together to achieve something similar on a domestic level, punctuated here by tulips rather than oak, ash, or hazel. There is wonder here, and the impulse to spectate as April turns to May and fern-leaved cow parsley rises from the woodland floor, the frothiness of Bishop's flower ascending in the flowerbed as lupin, paeony, and rose prepare to take the stage, is strongly felt. It seems there is no stopping the parade, each day bringing something new – wallflower, columbine, poppy, iris – the spectacle of a garden as spring reaches its apotheosis is hard to beat. Anyone, it is said, can have a garden that looks good at the height of spring, full and fresh and filled with flower but, while losing ourselves in admiration might feel an appropriate response to such vitality, the opportunity to join in is there to be seized.

Now the moment is right, while plants are in a phase of active growth, to get deeply involved, the consequences of our interventions quickly being revealed. Whether we gain more satisfaction from looking on as the spoils of garden centre raids quickly establish and grow, or from raising our own from seed or cuttings, the two approaches to introducing new plants are not mutually exclusive, and one doesn't necessarily make you a better gardener than the other. I love new things as much as the next person, no stranger to the thrill of building up a collection, but I wonder when we became persuaded of an obligation to add constantly to the stock of our borders with plants sourced beyond our own patch of ground, when we could work quite as well with what we've got, and leave the selection of new plants to the birds and animals who visit, and the seed bank in the soil. But the longer we spend in our own gardens, the more acquisitive we tend to get when visiting other people's and, while this is as true at any other time of year, all the seeds we may have snaffled, and cuttings we may surreptitiously have taken (it's amazing how "clumsy" gardeners are around other people's plants), will grow away with gusto now that spring is here. Those less inclined to minor larceny will shop from comprehensive lists made while browsing catalogues and surveying others' planting schemes, and whatever the state of the gardener's conscience, the result on the ground is much the same. Small plants become bigger plants, the rate of expansion rising by the day, the growth to which the garden bears witness in terms not just of stature, but in order of magnitude.

When there are so many hurdles for an infant plant to overcome, growing from seed is a high-stakes game in terms of emotional involvement, from cuttings marginally less so since much of the structure already exists (we'll visit both in *Growth and development*, see page 132), though still a process requiring time and attention. What if there were a way to catch a wave on this rising tide of vernal power – the warming soil still damp, the sun's energy brighter every day – to create more mature plants in an instant from those we already have, with not much more than a spade? Delightfully, there is – at least when it comes to those herbaceous perennials that tend to organize their underground parts as a tangled mass of wiry roots, rather than preferring one large, long taproot. Hardy geraniums take particularly well to this kind of treatment, as do globe thistles, heleniums, asters, and day lilies. A Siberian iris will love you for it. There is no complex alchemy here – at least, not on the part of the gardener – no fiendishly tricky technique to master, other than digging up a clump of the plant in question and chopping it into smaller pieces, each portion of soil and roots and nascent stem having the potential to grow away into a plant in its own right which, settled into the soil in its new location, it will proceed to do in short measure. Good form dictates the use of back-to-back garden forks to tease the clump into divisions, though you could use your spade at a push, and best practice would have you place your excavation on its side and operate laterally, rather than risking damage to the growing points by going in from above, but that's all there is to it.

Contrary to what might be expected, long-established perennials appreciate this kind of roughhousing, their long-standing growth pattern radiating ever outward and growing moribund in the centre and, maybe it's my age, but it's impossible for me not to feel the parallel. Winter was long, and surely we could all do with new beginnings, a sharp shock to cut away the old and set us back in place with renewed vigour and purpose. But nature is rapacious, and if plants make resolutions for the new year, they will unquestionably involve some kind of verdant jiggery-pokery that results in multiplication, and more of everything – more of the same, more of variety. More leaf, more flower, more growth, more green. More *life*… and they will start here and now, with the lengthening of days, and the coming of the light.

A free lunch

Life returns to the garden, a green tide rolling in, increasing in brightness and intensity as the days grow longer. This is no coincidence; light and growth are bound together through a superpower particular to all green plants, from the towering redwoods in the forests of Oregon to the lowliest dandelion in the suburban lawns of Orpington. The process, known as photosynthesis – really a series of chemical reactions – takes place in cells within the plant tissue rich in chlorophyll, the key pigment responsible for the green colour of stems and, especially, leaves. This is the crazy alchemy that not only makes your garden grow but underpins all life on Earth.

I've been out here for hours, constructing smart new bins for the compost. The construction is the easy bit, but first a temporary new home needs to be found for a tonne or so of partially composted matter, the remnants of rotten pallets used to give structure to the bays' previous incarnation hauled off to the bonfire (the untreated softwood of delivery pallets is great for use in short-term projects for the garden or allotment, but over the long term? Not so much). These tasks accomplished, there are barrow loads of perennial stinging nettle destined to become compost tea, the tracery of their saffron-yellow roots an indecipherable cartography through the soil. Delving deeper, brittle white bindweed roots laugh at every effort of manual extraction – I tease out the obvious bits and send them off with the nettles. There is some satisfaction in slowing the momentum of the bindweed's growth, but it is very much a part of the land on which the garden sits, and as such has a certain right to romp. All this stuff could be left in a heap to get on with composting in its own irregular fashion, but we fancy a little organization, and if we're kidding ourselves that some sturdy carpentry and a lick of paint will bring a modicum of bougie charm to this working part of the garden, we're still going to try. It's the kind of job that would benefit from the clarity and pace of winter, but ideally calls for the congenial ground conditions of those first few fine days of spring; warming activity that's not only thirsty work, but hungry, too and, while the Thermos flask and tea strainer have been pressed into regular service throughout the course of the morning (even in other people's gardens, tea must be done properly), I'm finding efforts to ignore the rumbling of my stomach increasingly distracting. Downing tools and peeling off gloves, I plunge a hand into the depths of my daysack, extracting the brown paper bag, bulging with promise. Inevitably, there is plant matter in my sandwiches, but not the kind of grassy fallout that I manage frequently to strim into a carelessly placed mug of tea. Here there are wheat and barley in the chunky doorstops of home-baked granary loaf, sunflowers in the dairy-free spread, corn starch and canola oil in the mayonnaise and a houmous of chickpeas, lemons, sesame seeds, and olives. There will be tomatoes and lettuce in the filling and, if today these are accompanied by a portion of meat or fish, it will be the mortal remains of some small soul who has, like me, been eating plants. Or who has eaten someone else who's been eating plants. Taking a break from the gardening, there's still no escaping plants.

I make quite a good sandwich. The trick, apart from fresh, preferably still slightly doughy bread, is generosity. Never stint on the contents, which should include fistfuls of leaves all bound together with lashings of gloop – mayonnaise, houmous, mustard, oil, and vinegar – though always in such a way as to avoid sogginess. I once found myself out of cranberry jelly when making a sandwich from the remains of a roast chicken I'd been sharing with Bill, and was forced to improvise by using jam. Both gardener and terrier, of course, declared the combination delicious; but it occurred to me that I had something in common with our erstwhile feathered friend, for whose sacrifice in the name of lunch I had murmured a grateful thanks. While I may be superior to a chicken in possessing the ability to make a sandwich, neither one of us, on our very best day, would be able to conjure the elements that go into our dinner from thin air, and particles of light. For that, we'd be needing a plant.

During the latter part of the eighteenth century, just as John Montagu, fourth Earl of Sandwich, was developing his eponymous creation at the gambling tables of the Hellfire Club, his contemporaries in the field of science were enduring a febrile phase in the advancement of their scholarly understanding, not least in that body of knowledge concerning the air we breathe. Though the name "*oxygène*" was first bestowed upon that essential portion of our atmosphere in 1777 by French chemist Antoine Lavoisier, we have British clergyman, chemist, and natural philosopher Joseph Priestley to thank for discovering the fascinating relationship between that gas and green plants. It was Priestley who, upon placing a lighted candle in a bell jar, noticed that while the air within became increasingly reluctant to support a flame, it could be "restored" by the introduction of living plant material. He records the outcome of one particular experiment carried out in the summer of 1771[9]:

"...I put a sprig of mint into a glass-jar, standing inverted in a vessel of water; [and] when it had continued growing there for some months, I found that the air would neither extinguish a candle, nor was it at all inconvenient to a mouse, which I put into it."

A happy outcome for the rodent in question and one that, while we're unable to comfort ourselves with the thought that candles alone were privy to the suffocating effects of the good reverend's experiments, clearly demonstrates the ability of plants to absorb that component of the atmosphere we now know to be carbon dioxide whilst at the same time emitting oxygen. Unwittingly, Priestley had discovered a key truth about the process of photosynthesis and, though it would be more than a hundred years before the term itself would be coined by the American botanist Charles Reid Barnes in 1893 – and longer, until it became widely used – it's no stretch to claim that the implications of his discovery hold the key to the nature of life as we know it.

For their optimum comfort and health, plants are no different from other organisms in demanding a comprehensive catalogue of nutrients to support their various bodily functions, but to carry out the most essential of these – to *grow* – they require energy. This they can create for themselves, using sunlight to transform carbon dioxide and water into the sugar glucose, a metabolic powerhouse that can be drawn upon immediately, converted into starch for storage and later use, or transformed into the cellulose needed by plant cells in the construction of their walls. As a by-product of the process, oxygen is expelled, as fortuitous for Priestley's mouse as it is for you and me, your grandmother, and the cat. But while photosynthetic ability is not the sole province of plants (certain bacteria and algae also have the knack), it does make plants fundamental to almost every ecosystem on the planet, the foundational layer of those hierarchies of consumption referred to by ecologists as "food chains" – familiar to every small child (albeit in an abstracted nursery-rhyme form), from the bizarre and macabre account of the old lady who swallowed a fly[10] – on which each successively higher level depends and without which the whole interrelated structure of life would collapse.

And the thing about this superpower, this ability of organisms classed as primary producers or autotrophs (literally "self-feeders") to whip up a meal from nothing more than light, air, and water, is that it's intrinsically linked to one of the properties of plants to which we most happily respond – their very *greenness*. "Greenness" isn't, you won't be surprised to learn, a scientific term, nor even the most elegant of words, but here it's a far more useful noun than "verdancy", if less poetic. Green is, in this sense, both the colour of a plant and the location where photosynthesis takes place; inside the pigment chlorophyll (from the Greek words *khloros* and *phyllon* meaning "green leaf") that gives plants their characteristic hue. We perceive colour as a result not so much of those parts of the light spectrum that an object absorbs, but those that they reject and reflect back to us. For reasons not yet fully understood, and rather like someone eating the bread from a sandwich before throwing away the filling, plants tend to bookend the available wavelength by using the red and blue ends of the spectrum while leaving out the green frequencies in the middle. Chlorophyll is not present in every section of the plant – you won't find any in roots or woody stems. Instead, this vital pigment is concentrated within microscopic organelles known as chloroplasts that exist within the cells of leaves, young stems, and various other aerial structures – those parts of a plant's anatomy most likely to be exposed to the sun's all-important rays. Because the first step in the process of creating your own food, it transpires, involves harnessing the energy potential of the sun to unlock some otherwise tricky chemistry, within the minuscule industrial complex of each chloroplast. Free to roam, in miniature form, through the chloroplast's double-walled interior, we would likely swim through a jelly-like matrix of proteins and nutrients, looking up at the grana towering overhead, stacks of disc-like structures (known as thylakoids) sucking in photons from cosmic rays that have travelled 93 million miles to reach them. In the centre of each disc there is a space in which the phase of photosynthesis known as the *light-dependent* reaction occurs, photons of light agitating the chlorophyll into such excitement that electrons are passed from one molecule to another. Water molecules are split apart, and energy-carrying compounds are passed out across the thylakoid membrane and into the jelly from which our imaginary observances are being made. And here, fuelled by this energy rather than directly by the sun, the *light-independent*, or *dark reaction* combines carbon dioxide and water to form sugar.

At any given moment of day or night, we are surrounded by this complex and essential activity, not only in the garden, but as we make our way through the country, the city streets, the office – anywhere there is living plant material. The reactions which together comprise photosynthesis have only very recently been recreated artificially in the laboratory, albeit with a far lower degree of efficiency than that managed by the scrappiest weed clinging on to life in a roadside gutter. As we as a species grapple with the technical challenges of capturing the sun's energy for direct and local use or for feeding it into our power grids for wider distribution, we have a lot of catching up to do with the plant kingdom, not only when it comes to the efficiency of energy conversion, but also in the construction of the necessary apparatus. While all green parts of the plant contain chlorophyll and are therefore able to photosynthesize, it is the leaf that represents the apogee of design when it comes to solar panels, not least through certain impressive auxiliary features, such as the ability to track the sun's movement across the sky, or to biodegrade entirely into valuable by-products at the end of their useful life.

Hold a green leaf in your hand. Twirl it around by its stalk – or petiole – or simply grip it gently between finger and thumb; familiar, reassuring. Simple. We made collages from such leaves for school projects as children (or when older, for posting on Instagram) and, even still, we shuffle through piles of them in autumn, or curse them in their soggy winter congregations as they block drains and gutters. When it comes to leaves, familiarity might not quite breed contempt, but it certainly clouds our appreciation of their importance to everything we hold dear. Look closer. When exactly was it that you or I first realized that a leaf has veins? Those same schooldays, surely; looking at them now, we're reminded of the ribbed tracery of a vaulted medieval undercroft, a structure of elegance and strength combined that somehow also calls to mind the network of vessels criss-crossing the backs of our hands. There's more here than ornament; function, too. But, while our reasoning might, through parallels drawn with our own anatomy, lead us to propose a theory of the transport of fluids essential to life running through this vascular network, we can be forgiven for failing to grasp the complexity of the work undertaken by what we hold, that the everyday object we turn over in our fingers is an organ that enables the plant to breathe, and provides it with food and energy.

Look more closely still, through a microscope now, at a section through the leaf. John Montagu's invention continues to pursue us through our garden musings, and here we find, sandwiched between waxy cuticles, a complex interior, organized in layers. The topmost of these, and closest to the sun, is a sheet of tightly packed cells that forms the upper epidermis, protecting the interior of the leaf from the outside environment. These cells exude a covering of the substance suberin, which helps to guard against the twin threats of uncontrolled moisture loss and the browsing of hungry creatures, while allowing the passage of light to the next layer, where columns of tall, thin cells packed with chloroplasts stand shoulder to shoulder. This – the palisade mesophyll layer – is the factory floor, where most of the plant's photosynthetic activity takes place. Beneath this, the spongy mesophyll functions as the leaf's concourse where, though the work rate is just as high, the prevailing energy is one of busy comings and goings. Here the cells, still containing chloroplasts, are much less closely packed together, allowing air spaces in between where the all-important business of gaseous exchange can take place, carbon dioxide and oxygen passing one another by in broadly opposite directions. This is the layer through which the veins run, bringing water and nutrients from the roots through the network of xylem tubes and ferrying away the products of photosynthesis to other parts of the plant in the phloem vessels. Supporting this layer and protecting it from the external world is another layer of epidermal cells along with its cuticle, in which pores, or stomates, open and close as necessary to allow the entry and exit of gases.

These stomates clearly play a critical role throughout the plant kingdom, and their ability to open and close – a routine, almost clockwork rhythm according to prevailing conditions in temperate climates, but a potentially fatal problem in arid environments – helps a plant balance its need for the carbon dioxide required for photosynthesis against the possibility of losing essential water to the atmosphere in warm weather. Where daytime temperatures could cause catastrophic water loss during the day, plants such as those in the stonecrop family Crassulaceae have developed the ability to keep their pores closed until the cooler temperatures of night, taking in carbon dioxide when transpiration levels are lowest and storing it within molecules of acid, before releasing it to the light-dependent phase of photosynthesis during the day. Though not restricted exclusively to the stonecrops, this adaptation is known as Crassulacean Acid Metabolism, with many CAM houseplants – including cactuses, orchids, the snake plant, and jade plant – being recommended for the bedroom for their ability to absorb carbon dioxide as we sleep.

Whether a plant chooses to open its doors by day or by night, this is the business of every leaf, from that soggy bit of lettuce you pull out of your lunch, through the basil that brings gentle heat and nose-tingling aromatics to your pasta, to those evil, stinging nettle leaves that, ripped off and boiled down, make as good a soup or pesto as any fancy herb nursed on the allotment or brought home from the market. The finer points of what goes on inside a leaf are clearly of great importance to commercial growers, who can manipulate and control the growth rate of the plants they produce by tweaking the availability of those ingredients essential to photosynthesis. Varying relative levels of water, carbon dioxide and light, along with the various trace elements needed in a supporting role – chief among these the mineral magnesium – is an essential skill for those who grow the plants with which we stock our flowerbeds and fill our bellies.

When the hurly-burly's done, what does a gardener need to know about photosynthesis, beyond what the word itself tells us, that the garden *makes itself with light*? An illuminating plant-care lesson pertains to the habit of variegated plants – those that have splashes of white or yellow on the foliage – to revert back to purely green should the gardener snooze over the removal of any non-variegated shoots, since it is these, with leaves of pure green and fully loaded with chlorophyll, that are able to outgrow their more ornamental, less photosynthetically-inclined compatriots. But, further than such esoteric insights, what does any of it really mean to you and me, standing here among the plants and looking about as the garden clothes itself in blossom and the leaves unfurl? Grateful as I am for my ability to articulate the equation of photosynthesis, to recognize an autotroph when I see one and appreciate the mechanisms within that allow it to create its own food, I'll sometimes allow myself to wonder what it is I've gained through book learning over and above what I might have gotten simply from experience and studious observation over time. It's hard to look back and unthink knowledge acquired, but I'm sure that, not so long ago, I just knew that as the strength of the sun increases, the garden leaps into life, that plants grow and that, being similarly solar powered, I felt a kinship with them; the same, but somehow different. There was wonder enough in this, if not so much in the way of understanding. Truth, we're reminded by the day, is the first casualty of war, but ignorance is cannon fodder against the onslaught of learning, and though ignorance and wonder aren't quite the same thing, aren't really the same thing at all, the one gets dragged down into the mud with the other. We find ourselves in the company of Lady Bracknell, pronouncing ignorance to be like a delicate exotic fruit: "touch it, and the bloom is gone."[11] It takes an effort of will, a certain confidence, even, to pick our childlike astonishment back up, dust it off and declare it still good for service – better even than good – and if understanding and learning look on with bitterness as we restore wonder to its rightful place, how much more indignant will they be when we place it not beside, but above them. Because wonder, given its due and tested in the flames of education, becomes a breathless kind of awe. And that, to me at least, seems an entirely appropriate state in which to find ourselves when entering the garden in spring, to watch it weave itself into being, with threads of bright light.

Growth and development

A garden never stays still – standing on a grassy path in spring, possibility manifests as a visceral experience. A faint prickling in the nose, a cool fresh taste of green on the air, filling the mouth and hitting the back of the throat, a hunger pang like a hole in the stomach. Something is happening out here, and we understand it in our bodies even if our brains take a while to catch on. For all the garden's ability to absorb us in the moment, so much of what we experience here is about the passage of time, of change and transformation. From the growth and development of individual plants to the maturing of beds and borders, all the time we're working on the garden, it's quietly working on us. But here, in this moment, it's all about potential, and this snapshot of time is filled with our anticipation of the garden to come.

Let's start at the very beginning, with the sowing of a seed, pausing long enough to acknowledge that in such simple acts we proclaim our belief in a future we hope to see, and our faith in a process we don't fully understand. No one expects a fanfare with each finger poked into damp compost, each seed settled into the shelter of the resulting depression, but there's satisfaction in knowing that with every seed sown, we declare our investment in all of our garden's tomorrows, and our intention to play a part for as long as we're allowed. This work is rhythmic and mindful, requiring no special equipment or paraphernalia, only a little water, some soil or sterile store-bought potting mix (to avoid the possibility of seeds already present outcompeting those we're sowing), and a packet of seeds. With experience, we may collect a few simple tools to assist our efforts; a selection of plant pots and seed trays, and a sieve or garden riddle to bring a happy consistency to the growing medium in which the seed will shortly come to nestle.

Within the business of gardening, the process of sowing seeds seems to be shrouded in mystery; when to start, how to prepare the seeds for the event, and whether they require light or dark conditions being a few of the topics hotly debated among aspiring horticulturalists. It's true that some seeds require more light than other, more bashful varieties that, presumably for reasons of modesty, would prefer a thin covering under which to make their first tentative forays into the wider world. On balance, most seeds prefer a little illumination in which to do their thing. It all seems overly complicated, especially when compared to those keen self-sowing perennials in our flowerbeds, or the weeds whose numbers increase by the day, with none of the attendant fuss. It's a slightly unfair comparison; the seeds sown by the garden are freshly fallen from the plant – by far the best way to do it – while the gardener keen to introduce plants from elsewhere often turns to seed that has been dried for storage and transport. Dehydration induces dormancy – and many plants will have in-built mechanisms built in to prevent them from germinating in an environmentally-unfavourable season. A thorough soak in water often provides sufficient reinvigoration, but those plants distinguished by a complex kind of dormancy in their seed require an apparently elaborate process of stratification; a warm treatment, followed by a spell of cold, moist conditions. That pioneer of scrubby woodland ground, the ash tree, spawns easily half a hundred copies of itself in my garden each year (and much to their chagrin, the gardens of my neighbours) with no assistance from anyone. And yet the rigmarole involved in inducing its saved seed to germinate reminds us that the key to sowing and growing is to mimic the conditions that would be experienced in its natural habitat, making the true value of the seed packet not the pretty picture on the front, but the sowing instructions on the back.

If this suggests there's a right way for every seed to be sown, we should be wary of extrapolating that into a general rule about gardening, where there's much to be said for simply giving things a go before watching and waiting to see what happens. The simple recipe of adding seeds to a tray of damp potting mix, and placing on a light, warm shelf could hardly be more straightforward and, if the result is other than anticipated, explorations can be made into what might be done differently next time. Whether that involves pre-treating the seeds by soaking or chilling them in the fridge or, once sown, by varying levels of light and heat, the perfect method can be arrived at through a process of trial and error, guided by research and conversations with other gardeners, and one route to the answer may be quite as good as another. But the key learning point here is that the seed – no matter how recalcitrant it might appear to be, whether it leaps into life with as little persuasion as mustard cress on damp paper towels or, like the ash, demands an absurdly baroque palaver before consenting to break its dormancy – wants to fulfil its destiny by growing into a plant. And the first step along that path is germination.

It is water, in the presence of oxygen, that begins this process: water that seeps through the seed coat or enters via the tiny micropyle aperture in the casing; water that swells the seed's cells and causes its cover to split; water that awakens the metabolism from slumber, hydrating the enzymes required to begin releasing energy from the reserves of stored food and sends the nascent root, or radicle, on a pioneering journey through the aforementioned micropyle, beyond the protective seed coat, and out into the surrounding soil.

A distinguishing factor among members of the plant kingdom is the number of embryonic leaves, or cotyledons, inside the seed casing. The seeds of a dicotyledon, such as a bean plant, pumpkin, or sunflower, will have a pair of seed leaves, familiar to us in the case of these edibles, and others like them, in the form of the plump half-ovals we might prize apart if we're making a particular meal of such nutty delicacies. So nutritious to us, these contain the energy-rich endosperm used to fuel the initial growth of the seedling plant. Monocotyledons such as rice, maize, or onions, have only one seed leaf, their endosperm separate from the cotyledon and remaining within the seed. In either case, as the radicle extends downwards into the soil and tentatively begins to establish a root system to feed and anchor the plant in place, the plumule extends in the opposite direction, elevating the structure that will give rise to the young plant's first "true leaves". Some plants raise chlorophyll-rich cotyledons above the surface of the soil to harvest the sun's rays, where they are often mistaken for the genuine foliage – at least until the seedling exhausts the carbohydrate within and they shrink away, allowing the unfurling solar-panels above to take on responsibility for energy supply. Not all seed leaves are hoisted aloft to cause such confusion. In the case of those plants that exhibit hypogeal germination, the cotyledons remain below the surface, and the first leaves to appear will be true leaves.

One seed leaf or two, above or below – by this point we've caught up with where we left off in the previous chapter, with plants using the energy of the sun to fund their growth. Here we are, still standing on the grassy path in spring and *feeling* all around the garden filling out and filling in. Other than in those speeded up time-lapse videos of pea shoots sprinting upwards out of their pots and whipping around until they find support, I've never seen a plant grow before my very eyes, but the change I perceive on each successive visit to the garden tells me that this must be happening. Something about growth never fails to engender in us delight, whether measured by the buckle holes of a puppy's collar, or the tell-tale hieroglyphics pencilled onto a wardrobe door, ostensibly charting a child's height but chronicling so much more. In the garden, we thrill along with the initial explosion of growth in spring, the bare soil and grey twigs that kept us company throughout winter transformed to living green in the shortest of times, and everything seeming to pulse with the energy and light that animates the contents of the beds and borders. This atmosphere continues until early summer when, showing little sign of easing off, the first hints of alarm intrude upon our thoughts as the prospect of unstoppable, Jack-and-the-Beanstalk-type expansion suggests itself, since by then we will have been reminded that a plant's growing ambitions are neither genteel, nor academic, but vigorous and uncompromisingly muscular.

The fastest growing plant on the planet is almost certainly a species of bamboo, with a recorded growth rate of almost 1 metre (3 ft) per day, or 4 cm (1.5 in) every hour. While *Guinness World Records* neglects to mention the species, two candidates would be madak or Japanese timber bamboo (*Phyllostachys bambusoides*) and moso or tortoiseshell bamboo (*Phyllostachys edulis*), each of which quickly achieve more than 20 metres (65 ft) in height. In my own garden, the average growth rate is far more sedate, though sometimes, in spring, it almost feels as though at any minute I might be tapped upon the shoulder by the tip of a thick, motile stem, requiring me to make way for its imminent progress. Daily in the borders, perennial plants are bulking up, geranium, aquilegia, cephalaria, echinops – great domes of foliage growing in density and circumference. Tulips, no longer politely pushing through the soil, flexing thick, strappy leaves, and extending chunky, bud-topped stems toward the sky. Crimson paeony buds, gaining momentum by the hour, with at this stage in their development an unfortunate resemblance to horror film hands, flayed and raw, thrusting upwards through the ground. Too fascinated to look away, I comfort myself with the anticipation of the short-lived but blousy beauty of the flowers to come, and the more enduring presence of their handsome foliage.

Surrounded by such vigorous industry and expansion, the garden appears to espouse a continual and steadily determined push from the moment winter departs to those last heady days of summer. Despite the daily increase in stem length and girth, foliage and ground cover, the promise – with the appearance of buds – of blossom, and the fulfilment of the same, this isn't exactly so. While not quite a case of "two steps forward and one step back", neither is it a continual straining, fit to bust. The growth of plant tissue is not characterized by barely sustainable exertion as a prelude to inevitable exhaustion, but by a rhythmic and determined surging forward – periods of growth with spells of recovery – and, though all of this is happening at a microscopic level and in the blink of an eye, it's another example of the garden offering instruction to those willing to hear. When it comes to stretching ourselves, we could learn a thing or two from plants.

At a cellular level, two things are going on inside the plant as it goes about the business of making more of itself. Individual cells create identical copies of themselves through a process of division, or mitosis, a process that occurs readily in the actively growing zones of plant tissue known as meristems, found at the tips of roots and shoots, around the nodes along the stem from which the leaves emerge, and just beneath the bark on woody plants. Toward the base of the meristems, cells are also able to grow in size and stature. It's this business of elongation that involves the twinned steps of relaxation, followed by stretch, an accomplishment in itself given that the walls of the cell are not entirely unlike the uncompromising walls of a house in their construction. Here, the basic building unit is not bricks of clay, but microfibril strands of cellulose, tied securely to one another by molecules of the complex carbohydrate xyloglucan and all bound within a mortar of pectin (that substance familiar to keen advocates of home-made jam in kitchens everywhere). What results is a strong structure with an inherent degree of flex, though not one immediately given to stretching overmuch – certainly not, without modification, enough to accommodate any significant expansion of its contents. Like a child's water bomb, the dimensions of each cell are largely determined by the pressure of the fluid it contains, though requiring a period of adjustment before they can expand; a pause during which their constituent parts uncouple, gather, and rearrange themselves in preparation for the imminent change. And so, with light, warmth, water, and nutrients in good supply, those hormones whose job it is to regulate plant growth step forward, activating the enzymes required to catalyze change in the fabric of the cell's wall, and causing brick and mortar to loosen and relax. The resulting baggy, flexible container now positively encourages the rushing in of more water than the cell previously held, which it does so enthusiastically, increasing the cell's internal pressure and causing it to stretch. This growth spurt now accomplished, the cell wall responds by laying down more cellulose to reinforce this larger iteration of itself, returning for the time being to a state of flexible stability.

Such flexibility, more than a mere physical property, seems to be a guiding factor in a plant's approach to the whole business of growing. Certain cells carry not only the genetic code for the entire plant but, under the right circumstances, the embryonic potential to create from themselves every specialized cell required by the adult plant. While the concept of such totipotency in animals – and especially humans – is the subject of scientific investigation and debate in the realm of stem-cell research, such a superpower has long been accepted to be pretty vanilla stuff for a plant, daily manufacturing one unspecialized cell after another in the meristems, each waiting to be instructed of its specific mission and identity, or consigned, in a state of spongy undecidedness, to a region where the bulk of this pith is needed to provide substance and structure. Mundane as this might be for plants, we have this phenomenon to thank for our ability to propagate them vegetatively, either through tissue culture, in which genetically identical copies of plants are grown in the lab from minuscule pieces of leaf, stem, or root, or from taking cuttings, in which the gardener chops a bit off one plant and sticks it in some potting mix to grow another. Most gardens being without sterile laboratory conditions, it's the latter of these two scenarios with which we'll occupy ourselves here.

*

Peer as closely as you like at the juvenile stem of a plant in spring; you'd never guess the nature of the activity within. But take another look, and let your gaze linger over the node, that point where stem and leaf connect. The arrangement of leaves follows a pattern, and most often they will emerge either in pairs on opposite sides of the stem, or singly, on alternate sides along its length. Sometimes, the foliage progresses up the stem in whorled spirals, but whether opposite, alternate, or whorled, the way in which a plant holds its leaves is used as an identifying feature, not as definitive as a fingerprint or a birthmark, but useful in differentiating one apparently similar plant from another. Such clues are of particular use in winter, when twigs, bare of leaf or bloom, offer few hints to a plant's identity, save for the colour of the bark and the arrangement of leaf nodes upon the stem. In that season, the brightly coloured stems of dogwood could be mistaken for willow, if it weren't for the fact that dogwood carries its leaves in opposite pairs, while willow's leaves are alternate (the initial of the appropriate arrangement handily provided by the second letter of their respective genera, *Cornus* and *Salix*). But beyond their appearance and arrangement along the stem, beyond even their capacity to produce the intricately complex photosynthetic apparatus we encountered previously, these swollen nodules speak to the gardener; a constant reminder not only of the potential to create more plants, but of the possibilities of a partnership between human and non-human custodians of this space.

Most often, we take cutting material by severing the stem just below a node. Stripping off any leaves save the topmost pair, we can plunge this propagule into potting mix, confident of having removed that tissue most likely to rot below the surface and create an inhospitable environment for our cutting to flourish. But if this fragment is to live and mature into an adult plant, we want it – unreasonably, you might think – to produce roots from portions of itself previously dedicated to doing nothing of the sort. The base of such a nodal cutting has two ways in which to oblige. Once more under the influence of hormones, latent root buds around the node awaken, while the unspecialized cells of the callus tissue forming around the cut wound receive instructions to produce adventitious roots. (Advice flip-flops regarding the need to dip the base of the cutting in rooting powder, which contains more of the same group of growth hormones – or auxins, named from the Greek word meaning "to increase" – responsible for promoting the plant's own rooting activity, but it rarely proves necessary in all but the most uncooperative of subjects.) Cuttings of certain plants – notably climbers, such as clematis – often fare better when the stem is severed midway between two leaf nodes, allowing the plant to shoot directly from the base when the cutting is pushed down into the soil with the lowermost node at the surface. This kind of "intermodal" cutting is more dependent on adventitious roots arising out of differentiated callus cells so, to increase their chances of success, some gardeners will create a longer wound along the length of the internode. Wherever the cut is placed, the ability of a stem to produce adventitious roots diminishes the more it nears maturity, especially regarding the production of flowers, for which reason cutting material is better taken from a non-flowering stem.

Faced with such an array of variables – cut placement, propagule preparation, and rooting location, to say nothing of the appropriate cocktail of water, light, and warmth required by these nascent plantlets – the absolute worst thing a gardener can do when taking cuttings is to lose more than a moment from those with which they have been allotted in worrying how it should be done. Some methods will be more successful than others, but by far the least effective is the one that begins with procrastination. That a plant wants to grow is a truism that often holds as well for its constituent parts as for the whole; we must trust to this innate vitality, ignore our own uncertainties, and let nature do the rest because, as any self-respecting plant cell knows, you have to relax and let your barriers down before you can grow. Though if we want to be found still standing from one season to the next, we may be required to acquire a thicker skin.

A thicker skin

A society grows great when old men plant trees in whose shade they shall never sit.
Anon

Children ask the best questions. As adults we can get a bit sniffy on being met with an enquiry such as "why are trees?" but, really, that's because ontology is something most of us waft through life doing our utmost to ignore. We might counter loftily with "why are trees what?" in order to buy a little time, as if we didn't know what our small interrogators were demanding, as if they didn't know we knew, only to be met with the clarification "why are trees *trees?*", landing us more or less right back where we started. It's a particularly fine question, and one most of us have no hope of complementing with a satisfactory answer. So, we deflect, and have a stab at answering another, such as, "what makes a tree a tree?" – a lesser enquiry perhaps, but an important one nonetheless, and one we have at least a decent chance of tackling without disappointing our audience.

The answer, of course, begins with *a big heart, and a skin thicker than an elephant's…*

Possibly, it's got something to do with insults. If we want to avoid being wounded anew with each critical comment or unkind observation launched in our direction, we're told to stop being so sensitive and to grow a thicker skin, the better to deflect the barbs with which life inevitably assails us. It wasn't until the vet diagnosed Bill with the liver failure that would eventually, and decisively, put an end to all his furry plant munching that I understood the word "insult" to have meaning beyond mere words. Within the medical world, it describes an acute and injurious event, often with potentially fatal consequences. Just the kind of thing you'd want to be protected from.

But there's also something about endurance going on here. Plants grown in an environment to which they're well suited do a pretty good job of shrugging off the attacks of pests and disease for a season. Those with an annual cycle need only be tough enough to withstand what spring and summer can throw at them and, if autumn is harder, once they've gone to seed, they're past caring. When winter comes around, they may sleep the sleep of the righteous on the compost heap, or as standing spectres in beds and borders. With more staying power, a herbaceous perennial abandons all top growth at the end of the growing season, spending the winter below ground, insulated by the soil on all sides, and covered over by a blanket of leaves or of mulch. But if a plant wants to maintain a presence above ground from one year to the next, it's going to need constant protection, not only through the milder days in the middle of the year, but from the harsh conditions at either end.

Is there also something in the answer about standing? Well, we'll get to the heart of that matter, too. In fact, that's just where we'll begin. Eventually.

In "On the Difficulty of Conjuring Up a Dryad"[12] the American novelist and poet Sylvia Plath laments the intransigence of an imagination too fickle to summon the nymph of the woods to the service of her poetry. Surrounded by the everyday clutter of the objects on her desk, she acknowledges her creative ambitions for the work in hand but, despite intentions to see her words take flight, they remain resolutely on the ground, and her trees stubbornly tree-like. I'm minded to learn from her experience, to avoid literary invocations of a metaphysical kind and stick safely within the realms of what I see before me, though my resolve doesn't last long. Because when it comes to developing a better understanding and appreciation of the garden around me, I find myself travelling an increasingly well-worn path to knock upon the door of my imaginative faculty. And so, in pursuit of an answer to this chapter's initial question, I feel the need to ask another.

Have you ever wondered what it would be like to be inside the trunk of a tree?

Perhaps imagination is unnecessary, and you have childhood memories of the hollowed-out interior of some venerable parkland oak, the perfect hiding place or den, with the homely roughness of the walls, the soft, trampled earth of the floor, a slight mushroomy smell and the oddly comforting knowledge of that bulk and weight above; a child, hiding behind the skirts of someone older, wiser, and eminently more capable of dealing with the world's uncertain challenges. Those of us with less rural upbringings are likely only to have encountered such wondrous beings within fairytales but, wherever met, such a space is open to the elements, on one side at least, and anyone can stroll in or out at will. What we need now is enlightenment, rather than hide-and-seek; a journey through the trunk from heart to bark following the pattern of growth, undertaken in a spirit of curious observation, and resulting, at the conclusion of our outward burrowing, with a greater appreciation of what informs the staying power of woody perennials.

*

Stand with me here at the very core, the heart of the tree. As we travel, it will help to recall an image of a section through a cut log, ripples radiating out from a pebble cast on still water with, at the centremost point, a memory of the gangly sapling that once was. Here is the pith, the oldest part, the dark smudge at the hub of the wooden wheel and, constructed from long-dead, spongy packing cells, the first area that might dry out, crack, and decay. The memories of trees, it seems, are as vulnerable to time as our own. We move, and encounter a ring of darkish wood, followed by another yet darker – noticing that, just as a city map reduces buildings and roads to two dimensional shadows, the rings we see in section are convenient representations of towering, upright tubes. This bicolour band represents a season's growth, the first, lighter ring comprising early wood of fast-growing, thin-walled cells laid down under the bright skies of spring and summer. Late or autumn wood makes up the next layer, darker in hue, its thicker-walled cells smaller and more tightly packed, a reaction to the more restrained growth rate of the year past its zenith.

But this is growth on a different plane from that which we've encountered before. In previous chapters, concerned with height and length, we watched as root and stem lengthened and reached upward to the sky or deeper into the soil. Here, a sheath of generative tissue girdles last year's growth, creating new cells by the same process of mitotic division, before and behind, in rings about the circumference – in total, there are two, though at this point in our journey through the tree we've yet to meet anything but the evidence of the wood they've left in their wake. This secondary growth, or *thickening*, is the phenomenon responsible for the year-on-year increase in the girth of woody plants. Moving forward into a lighter band of spring wood again, the pattern continues in concentric rings for as many years as the tree stands, the weather for every year of its life recorded in the generosity, or otherwise, of each ring's width.

Since we departed from the central core of pith, we have journeyed through heart wood, the strong internal pillar giving support to the tree as it grows in height and stature. Moving outwards towards one of the lateral growing points of the tree, the colour lightens and we transition from the sturdy centre to softer sapwood. Despite differences in appearance and moisture content, both are comprised of dead xylem vessels, though those in the sapwood are functional, conducting moisture and nutrients upwards from the roots and accounting for the rushing sounds of water all about, accompanied by the rhythmic *pop* of air bubbles in the liquid columns. According to the United States Geological Survey, a mature oak tree will lose over 400 litres (105 US liquid gallons) of water a day to the atmosphere through its leaves[13], placing significant demands upon the internal piping that transports fluid from root to branch, and providing quite the symphony for those who have ears to hear.

At the outer limit of the burbling sapwood we find the vascular cambium, the first of the lateral meristems. Here, at last, is animation, activity beyond the sucking and cracking of giant straws – the emergence of new, living tissue, its function clearly defined upon each side. The short-lived xylem tubes we first pass through quickly surrender a hold on life, consuming their own contents in the process to free the space needed for the translocation of fluid around the plant; a system of vessels at once dead, and still entirely vital. Beyond these selfless units, a cordon of duplication separates one zone from the next, an apparently unceasing conveyor belt in which cells split and reinvent themselves in their own image, on repeat. Ahead of this genesis band the phloem cells emerge into the cortex – the living half of the vascular equation, circulating sugary sap from leaf-laden branches to deliver energy shots where needed – would it be rude to take a sip as we travel? Still, we move outwards, and we have left the wood behind, in front now only bark. More complex and layered than generally imagined – thicker, too, encompassing everything from the outer edge of the vascular cambium to the open air, this fraction of the trunk's total diameter is the part of the tree that's most alive, a critical sheath whose integrity, should it be ringed entirely by rabbit, deer, or by gardener's negligence with strangulating ties or nylon-wire strimmer, will dictate matters of life and death.

And now the first hint of the nearness of light; neither glimmer nor spark, but the scattering of chloroplasts in this thin, loosely-packed ring of cells – the phelloderm – that doubles as both larder and water carrier. We stumble into the heart of another generative band, where cork cambium – or phellogen – throws out cells before and behind in the now familiar manner, creating in front the outermost layer of the bark, which marks our greatly anticipated return to the garden.

Feet on firm ground again, we turn and survey the rind of the tree, a corky layer known to botanists as phellem, which you and I will doubtless continue stubbornly to refer to as bark, despite now knowing better. Variable in appearance from one tree to another and, like us, smooth in youth yet wrinkled and fissured with age, the cork cells are tightly packed together, impermeable to water and mainly dead. If the wood of a tree can be thought of as a resting place for xylem vessels of years gone by, then the bark is where old phloem goes to die. But the cork isn't simply a hangout for has-beens. It has a critical function in protecting the living part of the tree from both water loss and attack by pathogens and browsing herbivores. Two substances are critical to this role – the first of these, known to every tea drinker and connoisseur of red wine in the drying of the mouth and a lingering presence on the teeth, being the tannins that guard against bacterial and fungal infection. The second, suberin, we met in *A free lunch* (see page 123) where it formed the waterproof coating to the epidermal surfaces of leaves. There, that waxy cuticle was interrupted by pores allowing the traffic of gases in and out, but while these stomata, situated in living tissue, can open and close according to environmental signals, the pores within the corky outer layers of a tree remain always open. These lenticels often form characteristic marks upon the bark, a peculiar kind of morse code that stands out upon the trunks of cherry, plum, or birch.

But though lenticels permit the passage of gas, the cork remains impermeable to water, largely due to the action of suberin, whose liquid-resistant properties make this layer of bark such a perfect source of material for the stopping of wine bottles. Suberin got its name from the cork oak, *Quercus suber* (*suber* meaning "cork" in Latin) – endemic to the Mediterranean and North Africa though most abundant in Portugal, where a generous layer of corky bark protects the trunk from seasonal forest fires. Increasingly planted for the ornamental qualities of its characterful and deeply fissured skin, the economic value of the cork oak is central to the regions in which it grows, its product being used in the manufacture of everything from cricket balls to drinks coasters, soundproofing to shuttlecocks, though the increasing use of plastic stoppers and screw caps in the wine industry presents a threat to its continued cultivation. Harvesting of the cork, strictly managed to avoid weakening the trees, is done exclusively by hand and, to ensure the continuity of the tree's vascular system, never involves the removal of more than a third of the cork layer and its cambium. It's a process that will continue at intervals of eight or nine years for half the life of the tree which, while not possessing the longevity of an English oak, will still get to enjoy a good 150 years in retirement.

What of other substances that make a tree a tree? Cellulose, the chief constituent of plant cell walls, we have also met previously, and this trees have in common with all plants, though here at the cork layer cellulose is responsible for only around a tenth of the total mass. But the woodiness of trunk and branch can be attributed largely to the presence of lignin which, while also present in herbaceous plants, is responsible for up to 30 per cent of the weight of wood. This resinous, aromatic material – presently the subject of research as a possible alternative to petroleum, not merely for biofuels but also in the manufacture of plastics – strengthens not only the cork but the walls of xylem vessels, where it bonds with cellulose to form a tough polymer with antibacterial and hydrophobic properties. This water resistance not only confers protection against rotting but reveals another of lignin's functions within the cells of the xylem, where its presence almost literally scares the sap water up and through the great heights and distances of the tree's veins. Despite its intractability – or maybe because of it, since nature loves a challenge – ligninolytic fungi have evolved to specialize in breaking down lignin into carbon dioxide and water, leaving decayed wood pale and fibrous behind. To these is credited the responsibility of white-rot in trees, while brown-rot fungi consume only the cellulose, leaving the wood brittle and dark in colour. The edible oyster mushroom *Pleurotus ostreatus* is a white-rot fungus, though its presence on trees is rarely parasitic, and more often indicative of other underlying health issues within the host. Far more aggressive is the pathogenic honey fungus *Armillaria mellea*, which can spread over long distances from the original site of infection by means of ropelike rhizomorphs, an adaptation that mimics the roots of higher plants. These long black "bootlaces", together with the white mycelial sheath found just beneath the bark of an infected tree and smelling strongly of mushrooms, are among the sights least welcomed by the gardener, though to the ecologist, they might be little more than signs of an organism looking for its dinner.

Albeit located for posterity somewhere in Greece, the originator of the familiar phrase at the head of this chapter remains cloaked in anonymity – frustratingly so, since it would be interesting to know why they felt it necessary to make the genial gardener both old and male. For the central truth here is that the life of a tree is no insignificant thing and, barring natural disasters and chainsaws, certain to outstrip that of any of ours. If, as has often been said, it takes the English oak 300 years to grow, 300 years to live, and 300 years to die, even the ash, one of its least long-lived neighbours in the wood, is likely to endure two or three times the span of our three-score years and ten. Though a silver birch might only be able to keep pace with the passage of our days, it would not be unheard of were it to linger for 140 years. Even a fleeting stay on this earth by the standards of a tree could number several of our own lives. Few of us will be so lucky to have such old trees in our garden, but even a young tree is transformative of the space, if for no other reason than being what the ground ultimately aspires to grow. But, so much more than this, here is a presence that will, along with us, see the years come and go while growing taller, and greyer, and thicker round the middle, speaking to us deep within, in a place where memory and remembrance reside.

Here is company. Every garden should have a tree.

Among the Indigenous peoples of North America, there are those who refer to trees as "the standing people" – they who watch us come and go, one after another. Placing an outstretched palm against the deeply diamond-hatched bark of an old goat willow or the spiralling creases of a tall chestnut, a certain kinship seems hard to deny, though the nature of the connection is beyond my ability to fathom, much less to articulate. But it's hard not to feel drawn to longevity, and there's something here I find worthy of trust, of protection, of *attention*, beyond any kind of environmental service such presence might perform, or any benefit I might gain in terms of ornament or function. What makes a tree a tree? Undeniably it's wood and bark, which allow it to endure, since endurance is key to the *treeness* of a tree – and these are the things that we can touch and feel and analyse. Less tangible, but no less real, it is a gentle kind of wisdom, and spirit, and that perennial presence that gives us the confidence to grow up in its shelter, with the guarantee of refuge in its shade.

To raise a plant

Train up a child in the way he should go, and when he is old, he will not depart from it.
Proverbs 22:6

Gardens don't stand still. This is where the fashionable preoccupation with treating the garden as "the outdoor room" and "extending the inside out" is always going to come unstuck. Gardening is more than just interior decorating *en plein air,* for the simple reason that most things inside the house (and here I'm not including the remote control, the LEGO you keep treading on, or your partner's shoes) stay where you put them. The plants in your garden don't. Moreover, while you'd be surprised if – other than a little wear and tear – your sofa didn't look pretty much the same from one month to the next, you wouldn't expect the same to be true of your plants.

For some, this is a source of frustration – the moment you get the place looking just so, it goes all blurry at the edges and begins to move about. Plants have an inexorable habit of growing, in the contemplation of which we've spent the last couple of chapters; this is the joy of gardening, and you can either get stuck in and get involved or stand back as they wind their way up and around you. But it's in that taking part – that moment in which we step forward and cease to be a bystander, when we stop looking upon plants as so much green furniture to be arranged about the space, where we stoop closer to the ground to check on the progress of a penstemon and, in so doing, give in to the temptations of symmetry by pinching one long shoot off between thumb and forefinger – it's in small but significant acts such as these that our involvement deepens and develops in meaning, and we discover the delight of helping to raise a plant.

But if this sounds like an attractive proposition, before we align ourselves to the sentiment, it would be as well to be sure of just what we're getting ourselves into. While self-examination might be our first impulse – have we got what it takes? Are we willing to offer the commitment required? – a better starting point would be with a definition of terms. To which end, what exactly do I mean by "raising a plant", and how necessary is it that we should do this work *ourselves?*

Compared to raising puppies or children, infant plants are decidedly more capable of getting on with things without our assistance – unless we cut them off from their natural environment by, for example, sowing them in seed trays at inappropriate times of the year, and keeping them isolated from the weather and the soil that would naturally provide them with food and water while they grow. We are placing ourselves, in such cases, *in loco parentis*, a position which brings with it the duty to do the best we can for our charges, directing the pattern of their growth towards a fulfilling adult existence.

But without our intervention (and even with it), nature raises a plant after her own fashion, as she was doing long before we arrived on the scene, calling upon substances within the plant's body to assist in this formative process. They say it takes a village to raise a child, with friends and family, nannies, teachers, and animated cartoon pigs each exerting an influence over different areas of its upbringing, but the congregation of helpers intrinsic to a plant's development is no less numerous and varied. To the auxins we met in *Growth and development*, see page 132 (and will meet again in *A cut above*, page 170) plants add gibberellins and cytokinins, abscisic acid and ethylene – five types of hormone that, together with a host of supporting chemical compounds, respectively manage processes including stem elongation, cell division, water management, and ripening, thus governing the growth and development of a plant from seed to flowers and fruit. Nature has it in hand and, at the risk of repeating myself, your garden will get along quite well without you. If a seed falls to the soil upside-down or on its side, its emerging radicle will know how to grow downwards into the ground. When a seedling grows in a shady spot, it will seek out the brightest area it can reach, bending itself towards the light. The plant just knows, in each of these cases moving auxins – evenly distributed through plant cells under optimum growing conditions – into greater concentration within the cells furthest away from the desired direction of growth. The cells on the outer edge of the curve grow more rapidly than those upon the inside, which bends the stem or root accordingly. In neither case has a plant needed direction from you or me. Roots will seek out water and nutrients and the best mechanical purchase for anchorage, leaves will turn their faces to the sun, and flowers will put on their best show to attract the suit of any pollinator that might pass by all without reference to any human gardener.

Myriad ways exist in which a plant adapts to its environment and alters its pattern of growth, presenting us with evidence of nature raising plants without our involvement. But often plants share their environments with us and, though we might pass them by unnoticed – as we're rushing along a busy street, clambering out of the car to get to the supermarket, or checking the platform indicator for the details of our late-running train – it's difficult to argue convincingly that our presence won't at some point intersect with those of the plants, our own patterns of behaviour, however unwittingly, affecting how they grow.

We've spoken already of the importance of *purpose* and *intention* to the process of gardening. If the garden itself is a construct (and I think it might be, albeit a useful and highly beneficial one), then these are key to our understanding of what it might mean to garden, and certainly when we start to interrogate that meaning, we tend to approach the questioning from our own point of view. But surely there's another perspective that's every bit as valid, if harder to discern with any degree of confidence – that of the garden itself. I find myself wondering if it's possible, long before we've accepted the title of "gardener" for ourselves, that the garden feels itself gardened by our mere presence, watching this accidental gardener move about the space, imprinting themselves upon the ground with each step and creating new opportunities for growth, new spaces to colonize, new objects to be conquered and clambered over with every wheelie bin moved, every shed erected, or kid's climbing frame and slide introduced. What tiny living thing might find shelter in the shallow depression of our footsteps on the soft earth, what long-hidden seeds find their way – at last! – into the light of day by the unintentional churning of the soil as we go about our business? What shrub might find itself roughly pruned by the violent arrival of a football within the compass of its twiggy perimeter, the accidental removal of mature old growth providing all the space and encouragement needed for juvenile stems to bud and flourish? For all our framing of the natural world in terms of harmony, equilibrium, and balance, the garden thrives on change, and nature is an opportunist more than happy to take advantage of our unintentional blunderings, especially if they offer more light, space, or other elements necessary for growth.

If it is possible that we can help to raise a plant unwittingly, it's when we do so through an act of will that we open ourselves fully to the benefits that will flow our way from such connection. Reciprocity is not something most of us go looking for in the garden; such altruism rarely comes naturally, and the garden's apparent independence could discourage familiarity. If we begin to take an interest in raising a plant, we should understand that, given the right conditions, it will grow away with or without us – differently, maybe more scrappily of its own accord, and harder, too, in form and appearance perhaps conforming not quite so much to our tastes. But the more time we spend with plants, the more we understand that most will submit to our interference and allow themselves to be influenced. Eventually, the mutualism that builds over time between gardener and gardened informs and sustains the relationship on a level so visceral for us that to part company with a garden can bring about a very real sense of grief. Somehow, in the process of raising a plant, the gardener not only gains the satisfaction of knowing themselves to have been useful, but finds themselves on the receiving end of an outpouring of validation and meaningful experience frequently and loosely referred to under the umbrella of "mental health benefits", but which might more usefully be described in terms of joy, acceptance, peace, serenity, purpose, focus – the list as long and varied as the roll call of gardeners who can attest to its existence. More than simply swapping attention for enlightenment, we grow together. Perhaps the magic occurs in the knowledge that nature can do so much of her own accord but, when we bring ourselves fully and purposefully into engagement with the garden, she allows herself to be moulded by us, giving form to the expression of our desires.

But plants, like people, remain closely and soulfully bound to their home soil, and something happens when we grow a plant in anything other than an environment with which it might feel at one. It's more common than you might think, and can lead to frustration when the rhododendron we've planted on our chalky ground goes a sickly yellow, drops most of its leaves and refuses to flower, or when the tree fern that adorned the patio all summer long with its verdant exoticism refuses to burst into life the following spring, owing to our failure to provide adequate frost protection through the winter. But it can also lead to a closer relationship with plants that find themselves entirely dependent on us to create conditions more nearly approximating those in which they are programmed to flourish, when we might be called upon to provide a growing medium entirely different to our garden's soil, or the consistency of life under glass, sheltered from the vagaries of a temperate climate. We might grow in containers through choice – to cluster plants upon the patio or around a doorway – or through necessity – by virtue of having no access to the ground – but with no way for roots to go in search of water and nutrients beyond the bounds of the pot, the requirement for the gardener to provide both is absolute. The title of "plant parent" is an entirely accurate (if slightly mawkish) description bandied about those corners of social media where a healthy obsession with houseplants is *de rigueur*, and where most who take the mantle upon themselves have no alternative but to grow plants inside their apartment, or perhaps – for the fortunate – upon a balcony. The plants, careless of whatever socio-economic levers are being pulled to create the conditions in which their presence within our centrally heated interiors has suddenly become desirable once more, do their utmost to grow with whatever somatic resources they possess but, so far from home, look to us to make up the shortfall. We have made ourselves indispensable to them.

It's not like that for all of us. There are many who love their gardens, who, through lack of time or confidence or inclination, have yet to discover what it is to raise a plant. When it comes to those who take an active interest in their garden, there are decorators and doulas, the difference between them being largely one of attitude and focus, at least when it comes to the garden as a whole, if not the individual plants within. The decorator looks to their own needs and desires, curating a collection of plants to be arranged and manipulated in such a way as to provide a pleasing environment for themselves, their family, and friends. Meanwhile, the doula looks to the needs and desires of the one in their care, helping them to transition into the next phase of their existence and, in doing so, to achieve fulfilment. With humans, this phase is typically motherhood, but sometimes a doula will be engaged to come alongside a person as they approach death, and both life events are entirely appropriate within the setting of the garden, where creation, nurturing, death, and rebirth all play their part. Our gardening personalities won't always align entirely either with decorator or doula. Rather, we need to allow room for both to coexist, making space for spectacle, whilst appreciating the meaningfulness of a relationship with the trees, the plants and their flowers, the soil and all the life within.

There will inevitably be times when all we ask of the garden is a calming spot to enjoy a nice glass of wine after a long day, and if we can also bring ourselves to celebrate fecundity and development, growth and death and all the messiness of existence, we give the garden itself permission to do its work on us; not just to absorb our tired emotional states and soothe our trouble, but to refresh and energize; not merely to minister to us, but to invigorate us, too.

There exists no set passage through which we're introduced to the practice of raising plants. It can be a gift we're given in childhood that stays with us through our lives, or it can hit us out of nowhere when we're least expecting it. Sometimes, it sneaks up and catches us unaware, realization dawning that we've gotten used to the company of this plant or that, and that the day doesn't feel quite the same if we've not spent a few moments checking in on each other's progress. You might notice how the leaves on the straggly mint whose tips you twist off to adorn your summer evening mojito begin to grow with increased density, more closely packed and appearing at shorter intervals up the stem, and find yourself wondering if what's going on here has anything to do with how the box balls that used to be wafty and loose have firmed up with regular clipping into tight spheres you could rest a book on. You might begin to wonder if a piece of the plant which, now you remember, was quite small when you first brought it home but seems today to be trying to escape from the pot, might be separated off and planted elsewhere – but at that point the rum kicks in and horticultural speculation has to be postponed for the while. But the seeds of interest are sown, your curiosity is piqued and soon you find yourself "pinching out" the terminal buds at the top of the leggy, store-bought petunias in the window box to encourage the plant to produce more flowering sideshoots lower down, both prolonging and increasing your floral display. And then you're trying the same technique on cosmos and dahlias and the cheery red geraniums that those in the know insist you refer to by the name "pelargonium", and finding it works just as well – flowers for days, and all with a little pinching out of ends.

You wonder if you do the opposite, nipping out most of the sideshoots while tying one stem into a stick and cutting off all the others, if you can grow that abandoned, corner-dwelling fuchsia that's been languishing in a sorry looking plastic pot into some glorious, lollipop-headed specimen because, really, what have you got to lose? And suddenly, every plant you look at seems to offer a chance for you to become involved with its development – *hey, I'm growing fine here, but you wanna come play?*

And maybe that sounds a little frightening; maybe it's overwhelming to be presented with such a variety of opportunity whenever we're in the company of plants. But the life of the plants in our gardens has great depth, and we can choose when and to what degree we become involved. To a point, the answer to the question "how much do we need to do?" remains, as we've previously explored, "nothing, really" – since, if you're fortunate to have access to a piece of ground, nature can get on with raising plants herself. But if we want access to this extra dimension of relationship and reward, we'll need to begin to explore the many different ways in which to be a good plant parent or gardener; fostering an environment shaped by a willingness to nurture, to take notice, to give time and attention and to provide for a plant's needs will provide a firm foundation. It might involve the sowing of seeds, almost certainly the taking of cuttings. Pruning, whether in the form of clipping, or pinching out, or something more artful, is likely to happen; watering and feeding, definitely.

"All this?" you might protest. "For every plant? Who has the time?" You'd not be wrong. But no one, however solitary, gardens on their own, and your elemental partner has your back while you're learning to pay attention. It may be that you need just the one plant with which to be in such close relationship. It will change how you relate to every other, in your garden and beyond.

How to SOW A SEED

The wonder of a seed is that we can understand it. Almost. A seed is probably the smallest beginning that we can comfortably wrap our minds around, while turning it over between finger and thumb. And we all know the power of beginnings.

Good things come in small packages; seeds frequently in pass-the-parcel type layers, each seed itself a package – an embryonic plant, bundled with breakfast, lunch, and dinner and a complete set of instructions for life. Shop-bought or collected from the garden, an envelope is usually involved.

A pause. A breath. In the presence of such strong magic, this seems both fitting and wise.

Slit the envelope's flap with a knife, rather than tearing. We're not animals. Annoyingly, it's often only after an enthusiastic rip that we discover that half of any information printed on the packet is now in the bin. Pour the contents carefully out into your hand. If this turns out to be a smaller, foil package, tear a corner off this and decant the contents into your palm, the better to make their acquaintance. Take a long, admiring look at the impossibly small capsules from which plants will emerge. Now put them safely to one side.

Fill a container with finely-sieved potting mix, and gently firm it down using a flat-bottomed object – a block of wood or a purpose-made compost tamper. Water the potting mix from a can with a fine spray. Importantly, before sowing the seeds, to avoid washing them out of position.

Spend a moment imagining how and when the seed might like to be sown, in the certain knowledge that being dried out, bagged up and sat in a succession of warehouses, transport containers, and garden centres was never part of the plan.

Sow your seeds. A plant does not wait for someone to poke holes in the soil with a pencil or dibber, but lets its seeds fall where they will. You may wish to sow into regularly spaced, gentle impressions (a pencil makes the perfect implement) – it helps to know where to expect a seedling to appear. Cover the seed lightly with potting mix (no deeper than three times its height).

The only persuasion a seed usually needs to permit the emergence of its contents into the world is a little water, but understand that the artificial conditions you are providing may engender reluctance. Many recalcitrant seeds can be made cooperative with an overnight soaking or, if they're large enough, by gently nicking the outer layer with a sharp knife. Some plants – larkspur, for example – appreciate chilling in the garden over winter prior to germination, but will settle for a fortnight in the fridge.

Place seed tray or pot somewhere with good light, where you will remember to water them daily. Emergent seedlings enjoy company, attention, and a regular drink.

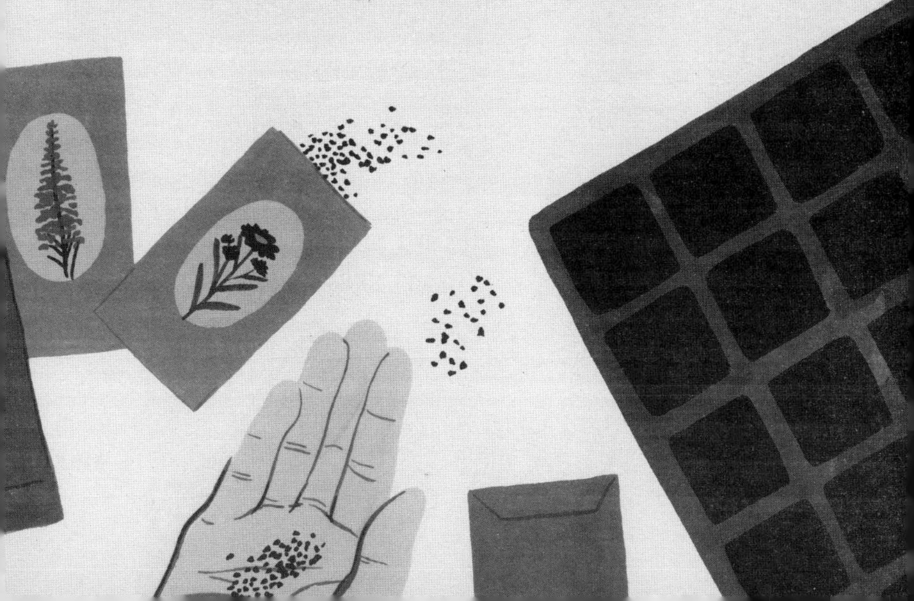

Picture perfect

Ah, but a man's reach should exceed his grasp,
Or what's a heaven for?
Robert Browning, "Andrea del Sarto" (1855)

"I just want to dig my whole garden up and throw it in the bin."

I'm frantically DM-ing Lucy, in that way you do when you suddenly realize you have a few precious moments of your friend's attention as she takes a micro-break from juggling "Unexpectedly Tricky Baby Number Two", the one that came into the world roaring and has yet to stop. The whole household is sleep deprived, and I can sense her frustration.

"...and I'm pulling out the bloody lettuces. They taste disgusting anyway."

Lucy's no quitter. She's a smart, sassy ex travel journo with a popular podcast and a string of online courses, relentless in everything and ruthless about reimagining any aspect of her life that doesn't quite work. But her garden is defeating her.

Anyone with the courage or foolhardiness to embark upon creating a garden for themselves must at some point work through a familiar dilemma. A dissonance common to many areas of modern life, a kind of Instagram-versus-reality disparity which exists as a shorthand to extract a kind of wry, companionable humour from the contrast between the impeccably styled image, and our un-cropped, unfiltered reality. We all do it, concerned – if we're honest – not so much with pulling the wool over each other's eyes, but with reassuring ourselves that we know where we're going, even if we're not quite sure how to get there. Life is easier to control and direct in these small, glowing squares, and a garden that fits into such a space remains as polished and tractable as on the day it was posted up for all the world to see. Splashing the details of domestic life across social media isn't to everyone's taste, but that doesn't preclude us from cranking up the kerb appeal of our house, fashioning a well curated and edited version of our gardening prowess to any who might pass by, with hanging baskets generously filled to overflowing, burgeoning boxes on every windowsill and the front step polished to a high shine. Perhaps everything in the back garden is primped to a similar degree, but there we might allow ourselves a little leeway in the areas only friends and family get to see. Whatever version of reality we choose to display to the world, most of us need a little assistance in bridging the gap between our notions of the picture-perfect garden on one side, and the reality of the space we find beyond the back door; whether that's a jumbled mash-up of wheelie bin, washing line, and bicycles accessorized with containers of the hopeless and dead and the half dead, or something with more order and visual appeal, but still some way away from the garden of our dreams.

Our gardens, however, will do their own thing and grow their own way. They don't read the same magazines, watch the same tinkly television programmes or follow the same influencers as you or I and, constantly in a state of trying to get back to that happy state of wilderness from which human activity has temporarily brought about the landscape's exile, their energies require redirection if there's to be hope of achieving anything resembling the images we carry in our heads. But depending on how they were received, these images themselves can be as pernicious as we have been led to believe are the weeds in our lawns and borders, whose constant presence prevents us from creating our own version of garden nirvana. We know that pictures in magazines and their online equivalents are designed to accentuate our own dissatisfaction with the stock of goods and chattels over which we claim ownership, though it's a crumb of knowledge that rarely prevents us from falling for the lure. Should it be different with gardening? We might like to think so, but whether editorial or advertisement, and irrespective of the quality of journalism, it all amounts to the same. These pictures encourage consumption but are rarely of help to our gardening ambitions in anything other than the short to medium term – a new barbecue or set of patio furniture is not suddenly going to transform a previously unloved outdoor space into a luxuriant and relaxing haven unless accompanied by some domestic behavioural changes. Of greater significance to us all are implications for the environment through our continued and apparently willing involvement in a system reliant upon harmful industrial practices, including manufacturing, transport, and the use of finite resources for energy and raw materials.

But if buying *stuff* isn't the answer, paying for services – should we be in the privileged position to do so – offers a more benign progression towards our gardening goals. By supporting skilled professionals within local economies, we enlist the opinions and experience of garden designers, landscapers, and gardeners, each of them bringing their own unique perspective and expertise to bear upon the land, and how it should answer to the needs of the house to which it finds itself in service. When it comes to increasing the stock of plants in our gardens, we can support growers too, an act of consumerism that makes its mark somewhere on the line between buying stuff and buying services, but rather more towards the latter, especially when plants are bought from one of the many local independent nurseries, whose dedicated staff are invariably only too happy to share their encyclopedic knowledge – unsung horticultural heroes, who can also often be found behind stands at one of the many plant fairs that take place around the country.

According to personal inclination, if not necessarily to values, I find I'm rarely an advocate for grubbing everything out of a garden and starting from scratch. It's less of a make-do-and-mend mindset than a fascination with the stories told by objects that have borne witness to the passage of time, coupled with the ability of well-conceived planting to transform a space; though I enjoy being challenged by those who have the vision to take a more sweeping approach and deploy hard landscaping with sympathy and understanding. Notwithstanding such a disposition, whenever we're asked to consider a garden in terms of how it speaks to the needs of the gardener it's evident that there exists some clear space between being grateful for what we have and being content with how it is; we need to get comfortable with the desire to make the garden work for us. When moving into a new home, it often takes months, even years, before we realize that half of our dissatisfaction with the outdoor space is attributable to us having been quite literally living with someone else's garden all this time – a space laid out to a plan that makes no reference to how we live or move around, and planted according to the taste and resources of a former owner or, in the case of new builds, the economic priorities of the housebuilder. There's no reason why we shouldn't strive to make the best of what we have, which is not quite the same as saying we should get rid of everything a garden contains and start from scratch. Despite the picture-perfect garden in our mind's eye, there's often something intrinsic to this spot that drew us here initially, and we're well advised to pause for thought before creating a real-life facsimile of our aspirational image. Because even while we're attempting to make positive changes that we think will bring about the kind of harmony we seek, there's another way for someone else's garden to sneak up on us that has nothing to do with the lumpy, green baize rolled out across a sea of rubble and compacted soil, or the narrow flowerbeds timidly clinging to the foot of the fences on each side with a rectangle of lawn bisected by a straight, concrete path.

Let's say you do transform your garden into the one in the picture. With no expense spared, every detail has been crisply executed, every crumb of soil swept from the patio, every plant primped to perfection. It looks flawless – but is it now, as your garden, perfect? Does it fit you like a well-tailored outfit, one that manages to combine gasp-inducing glamour with comfort, practicality, and more useable pockets than you could ever possibly need? It's unlikely, for the sole reason that you've been using as your guiding principle a picture of – once more for those in the back – someone else's garden. In this imaginary enterprise, even assuming no fault could be found in the implementation, the pattern was a poor fit; all your effort expended in recreating another's idea of perfection rather than deciding for yourself how you want your garden to be and investing your energies in creating that bespoke solution. And if, after all your efforts to find the right plants and source the correct furniture, the garden still feels no more comfortable than a showroom and bears all the relevance to the way you live your life that could be expected of such a space, this doesn't mean there's anything wrong with your taste, budget, or ability to chase down what you thought you wanted. You just need to change the picture for one of your own making. It won't be quite as glossy – more of a photofit, a composite of pages cut from magazines, images, and articles captured to Pinterest boards, and photos snapped while wandering around other people's gardens. It doesn't have to be pretty – most of us, unless we're professional designers or visualizers, lack the skills to create a beautifully rendered version of a garden that's yet to exist – but it does have to be lived with for a while, each element interrogated on a regular basis to ensure it continues to earn its place in the revised garden scheme.

For all its motley assembly, a "picture" so constructed quickly outstrips a simple photograph, however glossily appealing, in terms both of scope and utility. It's a metaphor for a set of disparate elements and ideas that appeal to us, with accompanying justifications as to why they deserve to be given space, and how their incorporation will improve, enhance, and reflect the lives of those who use the garden daily, and those who will be welcomed into it. And, unlike a static image that captures a moment in time, it is never quite finished, shifting subtly in response to the changing needs and experiences of the gardener, but always – in the case of the best ones – with reference to a common thread woven into every aspect, from plants to paths to seating and surfaces, bird feeders and ponds to trampolines and sheds. A garden designer might seek to tease out this kind of through-line from their client, the better to inform the brief and, since strands of commonality are often made out more clearly from a little distance, it can be easier for a third party – particularly one trained in such methods of detection – to pick them out. But if, when working on our own, our variously gathered inspirations don't immediately suggest common themes, it's best not to labour their identification for fear of forcing inappropriate restrictions upon our plans. In no time at all, our collection of ideas and elements will coalesce into a whole that expresses its own aggregate personality, against which any future additions can be judged for fit.

But there's something of the magpie about the way in which we go about consuming our garden influences, marking out with a beady eye the appealingly shiny and colourful and, if not actually breaking bits off and taking them back home to form a part of our treasure trove, taking a mental note of those plants or ideas that we intend to steal for ourselves. We probably spend most of our time thinking about how we'd like things to look in the garden and, while this is clearly a key consideration, it's not the only one. There's another picture that can help us here, but it's not something you can hold, stick in a photo album, or even save to your camera roll. To make the best use of this picture, you won't even want your eyes open.

*

So, step a little way into the garden with me again, and stand with your back to the door through which you just left the house. With eyes closed, it's easier to see the garden as we want it to be. We need to immerse ourselves in this space for a few minutes, to breathe it deeply in; soil, stone, tree, bird, sky – *we are in the garden, and the garden is in us*, and what a lot of rot that sounds but still, we'll give space to the voice that thinks this so and thank it for its insight, take another breath and feel the reassurance that – actually – yes, we were right. We thought we were. We are in the garden, and the garden is in us. The details that distract us daily from this truth are elsewhere, and we need not concern ourselves with them just now. And if they present themselves before the mind's eye as we stand – plants, path, furniture – we can swipe them gently but resolutely aside. That uninspiring fence – *swipe*. The tangle of honeysuckle that gets blackfly the moment it flowers, the patio that doesn't make the most of the evening sun – *swipe, swipe*. The best days you spend out here – on your own, with family, with friends – have nothing to do with things, and everything to do with experiences. And so, we focus not so much on objects and features, but on feelings – how do we want the garden to feel? Specifically, how do we want to feel while we're in the garden? Energized, calm, relaxed, excited, happy, peaceful, sociable – we ask our gardens to play many roles and hold us in different spaces, often simultaneously. Details can be dealt with in good time, but for now, we can stand with these feelings, conjuring the situations in which each might arise, imagining memories yet to be made.

*

I catch up with Lucy a few months later and find that, typically, she's pushed through her frustration and is making progress with her garden. She got rid of an outsized shed and used the space for a small lawn for the kids to play on. She surrendered the lettuces to the snails – who found them infinitely more appetizing than the family did – discovering, in doing so, her daughter's interest in the garden's wildlife, to the point where snail feeding time has become a regular part of the day. Following a crash course from her mother, she's forming opinions around weeding, responsibly eschewing both noxious chemicals and weed burners in favour of a patio knife with an L-shaped blade, perfect for her courtyard garden where the weeds love nothing better than winkling themselves into vanishingly thin gaps between small, square paving setts. The garden is a work in progress, but one where the family is happy to relax, and the children more than content to lose themselves for half an hour playing with mud pies in a big flowerpot. And all of this, despite her initial reservations. "It's intimidating," she tells me. "I find gardening to be intimidating," with which confession she's by no means alone. Talking with her further, I realize much of the blame for this lies with gardeners themselves – gardeners like me, and possibly even gardeners like you – partly for the way we can thoughtlessly bandy about impenetrable jargon, but partly too for setting unrealistic expectations with a refusal to share the stories of anything other than our gardening successes. For drawing a veil over the slug-munched, squirrel-dug, crinkle-leaved realities of sharing the garden with wildlife and multiple calls upon our time and attention, in favour of a ruthlessly edited highlight reel. For displaying images of our brightest and best as if these are the norm, and maybe for lacking the confidence to display the many inevitable failures along the way.

When it comes to gardens, there exists no Platonic ideal, some one-size-fits-all perfect state of garden-ness towards which all should aspire, for which the "picture perfect garden" would surely anyways be a poor substitute. It used to live in the pages of magazines, but in a world of unprecedented connectedness, we bear a collective responsibility for its dissemination. It rarely exists in reality and when it can be found, it lingers just as long as it takes to press a camera's shutter button. It isn't yours or mine – we know far too much about our gardens to believe the camera's lies. And yet even the enlightened can tumble into the comparison trap, to be filled with frustration when our own efforts fail to reach the dizzying heights of the standards that we imagine being set by others. But picture perfect is a myth and, like all legendary quests, the search for it ultimately both exhausting and fruitless. Far better to relax into life in the messy middle ground that lies between how we imagine we'd like the garden to be, and how it is just now – that gap of aspiration that finds us striving to make the garden better as we, in turn, become better gardeners. Understanding comes upon us with the acceptance of the truth that our garden will never be finished, and we ourselves are works in progress. And – quite unexpectedly – in that fit we find perfection.

How to WATER A PLANT

Plants, like people, need to maintain their levels of hydration – it's what keeps them looking plump and perky. Water pressure is what keeps them standing, while the wet stuff itself is an essential component for growth. Ideally, a plant would get all the water it needs from the soil and the weather. But a plant in a pot is entirely reliant for a drink on those who confined its roots, and even in the ground plants introduced to the garden are, at least initially, more dependent on us than their wilder, hardier neighbours with more established root systems and deeper local knowledge. But water is a precious resource and cannot be wasted. It's up to the gardener to make every drop count.

A pause. A breath. A moment for a thought.

The giving of sustenance is an act of love. Let this percolate through all your plant watering ministrations.

A can with a large, easy to fill opening, a long spout to reach through foliage, and a removable rose on the end to vary the spray, will serve both you and your plants well. From the rainwater butt, a jugful of old dish water or, if you must, from the mains tap, fill the can, all the while listening to its rising tone. Learn to recognize the pitch of the note it will sing when full. Turn off the tap. Pause. Take a breath.

We take our ability to carry the can for granted. Whether it contains half a litre or ten, we owe our bodies both gratitude and a moment's preparation.

Roll your shoulders, engage your core, bend your knees and, with a straight back, pick up the can, take it to your plant. Then stop. Look.

You can kill a plant with too much love, and roots will drown without access to air. Ensure that any excess water can soak into the soil or drain out of containers. Pots without drainage holes are inviting heartbreak.

If in doubt, water houseplants from the bottom. Place their porous inner pots into a basin filled with a little water and allow the roots to take up the moisture they require. Avoid dousing foliage, which can encourage fungal conditions, and aim for the roots.

Position the spout close to where the stem emerges from the soil, tip the can, and give your plant a long drink. Longer than that… a thorough soaking every couple of days is invariably better than a twice daily dribble.

Think of watering as an opportunity to spend time in the company of your plants; to check in, and see how they're holding themselves, and what they think of the weather. A drink with friends is always time well spent.

A cut above

I became a gardener at the very moment I discovered the truth about pruning. Until then, I had just been doing stuff in the garden. Like many, I had thought that secateurs and loppers were tools of last resort, to be brandished only when a plant began to outgrow its allotted space and very much with the intention of reimposing order – that despite all the sharp edges involved, pruning was really something of a blunt instrument, a way to stuff an over-enthusiastic plant back into its box. With that notion of control, I'd blundered unknowingly into a small corner of the whole picture, but there was one piece of information that would take my understanding of plants and gardening to a whole new level, and begin to suggest how a human being could work hand in hand with the forces of nature to create and shape a garden. And it all hinged upon one simple fact – that, contrary to what you might expect, cutting a plant actually *encourages* it to grow.

I suspect that Euripides, or whoever it was that first came up with the legend of the Lernaean Hydra before that Hellenistic gentleman felt called upon to record it for posterity, may have been a gardener. For as he tells it, whenever the hero Heracles cut off one of the fearsome creature's heads, two grew back in its place – a general principle not without its analogue in horticulture, albeit one attended to with more modest weaponry (chainsaws aside), and less chance of encountering lethally toxic sap. Of all the ways in which we attempt to understand plants by seeking out common ground with our vegetable kin – how we, too, need food and water to live, how we grow, and breathe, have sex, produce progeny; how we die and decompose into molecules of not dissimilar composition – there's at least this inescapable difference: that if you chop off any part more significant than nails or hair, the chances of you or I growing back a replacement (let alone two) are zilch. And so, we find something counter-intuitive, even intriguing, about the process; an operation that we might be forgiven for having considered to be all about subtraction turning out to be something quite different altogether, one where we begin to consider ways in which we might make creative introductions, working with a plant to decide, by mutual consent, how it will grow.

Removal, then, runs the risk of becoming the least most interesting thing about pruning, although used wisely it can contribute the light and air necessary for healthy and productive growth, freeing fruit from leafy obscurity to ripen in the rays of the sun while creating the kind of breezy thoroughfares amongst leaf and branch in which mildews and moulds find themselves disinclined to linger. Nonetheless, every pruning venture needs to begin in the same way with, before anything more ambitious can even be considered, the removal of anything we don't want, any stem, branch, or twig that could reasonably be described as "Dead, Damaged, or Diseased" being dismissed the instant it comes to our notice. And that notice is key, a degree of focus and attention required in assessing the plant before us, a visual checkup of its state of health. After a while, we begin to see not in terms of single units – tree, shrub, herbaceous perennial – but a complex collection of closely related, but disparate features – stem, branch, leaf, flower – with an underlying structure that might bear "improvement", beginning with the identification and removal of any tissue that could threaten our plant's ongoing health. Dead material is the easiest to spot, most often by its colour, since those parts from which the sap has retreated for longer than winter's duration will be wearing the habitual pallor of the departed; tones of grey, brown, and black in shades evolution has programmed us to recognize.

Snip.

Damage can arise from something as apparently benign as two stems rubbing together in the breeze, though stronger winds might wrest whole branches from a mature tree, leaving wounds whose jagged edges can be cut back to smooth tissue to aid the healing process. Wildlife, too, out of hunger or adolescent boredom, appears to think nothing of breaking a plant's skin, but why allow the patient to waste energy healing a damaged branch or stem when it could grow a healthy new one within the same time?

Snip.

Perhaps the signs of creeping disease are the hardest to spot, although fungal blooms on foliage, discoloured stems, and weeping cankers are all sure signs of afflictions that need cutting out or cutting off, in the hope of preserving the rest of the plant.

A *snip* here, as elsewhere, disinfecting tool blades with a mist of rubbing alcohol from a spray bottle kept to hand – you can never be too careful. It's mindful, engaging work, but not, you might argue, without a hint of the mundane. Were the Three Ds to describe the limit and scope of pruning, the business would have enough to recommend it and, although there's more potentially exciting stuff to come, this initial stage should be seen as so much more than a tiresome but necessary overture to the main event. This is triage; an act carried out by those who find themselves called to care – and for me, it's where relationships are made. Let me explain.

At some point quite soon after moving into our first house with a garden, I realized that I enjoyed the company of plants. Those who have been bitten by the gardening bug will recognize the symptoms of a growing obsession, from an inability to pass by any garden centre or roadside stall with so much as a potted primrose on display, to the growing collection of books, magazines, and seed catalogues and a tendency to give impromptu lectures on the topic of soil or compost or the germination of tomato seeds whenever the opportunity presents. But more than seeing the cultivation of a garden as a means to an end – to grow food and flowers, to spend time in the fresh air and the sun, to exercise that combination of creativity and organization that makes the best gardens come together – it became apparent that, in the presence of plants, I felt happier, more relaxed and, I came to understand, quite at home. And the opportunity to spend quality time with any individual, albeit one rooted to the spot and with sap running through their veins, has never yet failed to hollow out a space in the day where time somehow manages to pass with both agonizing torpor and unnerving haste, as might be expected when seen from the perspective of one who might have only a season or two in their current form, or decades, and whose toes are tickling around deep in the soil, knocking about with chunks of planetary importance.

When it comes to expressing our love for the natural world, some will throw their arms about the trunk of a tree and hug tightly. Others might choose to place a palm against the leathery bark to restore a connection we've disdainfully thrown aside. I understand the impulse, as well as why people are drawn to the largest and most long-lived members of the plant kingdom in order to tether themselves to something that seems to offer security and a route back to ways from which we've wandered so far. I hope I never lose a sense of awe at the majesty of a tree, but that feeling of groundedness comes to me, with quite the same cogency, through the rough leaves of creeping buttercup, or when raking my fingers through long evergreen grasses to comb out the dead foliage. Undoubtedly, trees are much longer-lived than the fleetingly herbaceous, but even trees will come and go while the soil, if we would only look after it, endures; anything growing in it, from the tiniest bittercress to the tallest redwood, embroidered with the same thread of eternity. Viewed through the prism of ages, the most mundane of garden tasks appears in its execution a vanishingly small price to pay for the moments spent in such company, and the act of triage begins to look less than entirely altruistic.

We have responsibilities here, too. If an inevitable consequence of pruning is growth, it's incumbent upon those who carry it out to ensure that the foundations for such expansion are as sound as can possibly be. And though, more often than not, a plant in the wild will overcome its injuries and insults with remarkable resourcefulness – sealing off damage, healing wounds, growing around the kind of obstacles and intrusions that would make a horticulturist flinch – a plant in the garden is not quite the same thing, being not only a representation of botanical resilience, but a reflection of the partnership that exists between human and non-human gardener. If we have a responsibility to the plant to carry out these most basic inquiries, we have just as much to ourselves, the better to ensure the garden provides the kind of testament to our efforts that we would wish for.

And if we are set upon this course of well-intentioned interference into how a plant should grow, we need to be on something more than mere nodding acquaintance with how it all goes together; how it branches and twists, how it yields to pressure, how it might be coerced into this shape or that, and the rules that govern the ways in which it might bend and snap. Beyond the immediate practicalities, beyond even the less concrete benefits to heart and soul, anyone who ever struggled with beginning a new project will appreciate a routine that instills an immediate familiarity with the raw material as, thrown together with barely an opportunity for introduction, we embark upon a drill that has value greater than its immediate horticultural benefit; as we make our careful progress around the plant, searching for signs of weakness and, where we find them, cutting them out. No decisions to be made, no arguments; follow the checklist along this stem, and move directly on to the next. And still there lingers a promise of the kind of creative partnership that had us reaching for our secateurs in the first place, eager to make our own mark upon what grows around our home.

Pruning occupies a central role in the health and beauty regime of the garden. Having dealt with the first concern we can proceed to the second, although beauty generally being accepted to reside in the eye of the beholder quite what this means for our garden is open to interpretation. Perhaps the size of a plant, or its form, or flowers, or fruit – each of these aspects can be manipulated to a greater or lesser extent with a little judicious surgery, although in every case, it's as well to have some understanding of the impact our cuts will have upon the patient.

A plant wants to grow; and grow according to the pattern and form of its genetic inheritance, though it reserves the right to make adjustments on the fly as dictated by environmental stress – a pronounced lean towards the light, an inclination to hug the ground in exposed and windy sites, even the propensity for the yew trees in our local wood to pull their roots out from between the huge sandstone boulders of their childhood home and wander about in search of richer soil. The freedom to choose such eccentricity seems baked deeply into the plant's psyche, but the level at which the gardener gets to exert some influence concerns the way the plant arranges its various parts, in particular bud, leaf, and stem – and it's all to do with a leafy plant's irresistible appetite for light.

All organisms are competitive, and plants are no exception, evolving a variety of different means to reach into the life-giving light and not being above using the neighbours for footstools. But whether they seek to grow faster than the rest, smother everything in their way or haul themselves up another individual by means of prickles deployed as grappling hooks, plants tend to prioritize their upward growth, quite literally over and above their outward expansion. It's a chemical mechanism controlled by the bud at the very end of the tallest and longest stem, which produces auxin hormones to inhibit competing growth. Cutting off the stem's tip puts an end to the tyranny of the apical bud, allowing dormant buds further down to spring into life and increasing the number of active side shoots. It's the reason we must endure the painful experience of cutting at least a third off the height of a newly-planted deciduous hedge, the better to avoid a bare-legged barrier in future years, and the mechanism that allowed me to balance a mug of tea on the box edging in *Ever green* (see page 100). It's the cause of us pinching out the tips of our dahlias at the beginning of May to give them the chance to grow into sturdy, garden-worthy plants; it's why we edit our pelargoniums and trim our lavender, creating the kind of ample silhouettes that speak to us of healthfulness and flourishing. It's doubly necessary with plants such as these from climates where a greater level of ambient light would be the norm (Mexico, the Antipodes, the Mediterranean) – under our gloomy skies any legginess caused by the development of one or two natural leading shoots being amplified by etiolation.

It's the secret behind every hedge and obsessively-clipped piece of topiary, from the privet elephants of Finsbury Park to the yew birds of Hidcote, the billowing, green clouds of box in the garden of Belgian landscape architect Jacques Wirtz in Antwerp, or the jostling together of neat but disparate shapes arranged by the French textile designer and stylist Nicole de Vésian in La Louve, Provence – not, in this case, the replacement of one stem with two, but the simple removal of a single bud, an event that sends a ripple cascading back downwards from amputated tip to origin, rousing every bud along the way from a long but hope-laden sleep. Putting to one side the intimidating volumes of advice on the subject and the concomitant concern about getting it wrong and sending your plants to the compost heap before their time, it's about little more than the knack of making a bush bushier; the joy of this being that, though all of this delicate snipping might sound like an onerous task across an entire garden, you don't have to painstakingly nip off the end of every shoot at the point of your secateurs. It happens by default as you're clipping a hedge, or shaping a shrub, or tackling the topiary peacock – apical dominance is defeated with every closing of the shears' handles as we assert ourselves, to whatever degree we feel appropriate, over the form and carriage of the plants under our care.

When we're clipping, what we're really doing is pruning in volume, a situation in which consideration of the precise placement of the cut on each stem would be impractical. The species and variety of plant chosen for hedges and topiary are deliberately selected not only for their visual appeal, but for their robustness and tolerance of such a necessarily casual approach, not least regarding the speed in which new growth draws a veil over any scars. But whether we're trimming off material over a wide area with hedge clippers or shears, or attempting a more surgical removal with secateurs, we're inviting a similar response from the plant, and so the issue arises of where we should place the cut when working with less sturdy specimens or those with larger leaves and clearer stems, where the effects of our industry are more readily discernable.

The Royal National Rose Society was once the world's oldest specialist plant society, founded in 1876 (the Royal Horticultural Society, which predates its inception by some 72 years, being instigated on the wings of pottery heir John Wedgwood's interest in all matters horticultural). It entered administration in 2017, citing a perceived decline in the popularity of the rose, though just a year or two later a surge of interest in the flower, driven by social media platforms such as Instagram and Pinterest, would surely have seen an upturn in its fortunes. In the 1990s, the society set up a test that would ensure furious debate within the rose-growing community for years to come, to say nothing of the barrage of unsolicited advice offered to any gardener engaged in a careful and considered process of tending to that particular shrub. Irrespective of the details of the test – its goals, conditions, and controls – the wider legacy of the study has been to introduce legitimacy to the notion of pruning roses with mechanical hedge-trimmers, suggesting no apparent ill-effects upon floral performance and results at least on a par with, if not surpassing "traditional" pruning methods. What this legacy overlooks is that, notwithstanding every plant's in-built drive to set seed and multiply, the long-term care regime cannot afford to elevate flowering over and above the preservation of a healthy plant in all its various parts. A letter from one of the original gardeners on the trial to the journal of the RHS[14] later revealed that those roses given a rough mechanical chop were more prone to disease, which began in the hedge trimmer beds before spreading to the control group. Demanding beauty at the expense of health is no basis upon which to build a relationship with our gardens.

A garden rose demands a little more cossetting than the wild cousin that weaves its way through hedgerow hawthorns, feted to by mowed into submission every summer by the brutal chomping of a tractor-mounted flail. More than being unsightly, leaving stubs of stem unadorned by foliage can be something of a health hazard to any plant, irrespective of location. No longer topped with a functioning bud and surplus to requirements, the redundant stump has its supply of water and nutrients shut off in short measure and, while the "dieback" theoretically stops at the next healthy bud, the progress of the necrotic tissue may continue downwards until cut out, with the potential to kill the entire stem. Placement is everything, and whether we're cutting a rose down to shin height or back to an established framework, whether it's tied into a wall or freestanding in the borders, arched over and tied to itself, we look to make any cut just above a leaf and sloping away from the bud, to direct any water away from where it might be tempted to linger and encourage rot. Such indulgence is possible with plants which, like the rose, have buds arranged alternately along the stem, but on a plant with opposite buds we cut straight across at a similar distance above the nodes. We keep our tools clean and sharp and our cuts neat and precise as a kindness and because, though nature might well be able to heal a ragged wound, why stress the plant any more than necessary, when that energy could be used for growing roots, or leaves, or stored up against flowers and fruit? These are the basic courtesies we can show any plant, and it calls for little more than a rag, some oil, and a sharpening stone.

Timing, though, if not quite everything, can play a significant role in the behaviour of the plant after pruning, and here it's best to pause, and ask ourselves quite what we're pruning for. Evergreen plants grown largely for form and volume can be cut at any time, except when the wounds we inflict or the soft, new foliage we encourage to grow might be vulnerable to extremes either of cold or heat. Deciduous trees and shrubs are often portrayed as most receptive to interference – whether that be pruning or planting – while dormant, but that's a terrible guide if you want to restrict growth (when you should be pruning in late summer, when they're in full leaf, see *Bare stems and skeletons*, page 78) or preserve the floral display of spring-flowering shrubs (which need pruning immediately after the flowers have faded, and then to be left alone). The list of rules for pruning can seem endless and confusing, with a different approach for every plant, but each rule is nothing more than a parcel of information based upon the experience of those who have gone before. They offer us a shortcut to the knowledge we would gain ourselves from close observation of the plants in our garden over time: whether a shrub produces its flowers on fresh new growth, or upon older wood produced in previous years; whether the mass of stems we think might benefit from being thinned out to admit more light and air are all the same, or if some bear the dried remnants of last year's blooms, while others, smoother, more vigorous, have yet to flower; whether this variety of apple seems inclined to flower and fruit on spurs along the length of each stem, which might encourage you to shorten stems and side shoots to encourage more spurs, or only right at the tip, which would persuade you to leave the terminal bud in place. Observation and experience are hard to beat, but in the absence of inordinate longevity, rules and advice can be a good substitute. There's only so many plants one gardener can get to know well over the course of a life.

It seems complex because plants are not simply plants, and pruning is not clear cut. There's no one-size-fits-all solution for chopping bits off the things and getting them to behave according to our desires – each is an individual, deserving of our time and attention. But whether we work out from first principles a pruning regime that suits both their needs and our own, or glean the same from books or YouTube, any reticence we might feel about this level of interaction with our garden arises out of fear rather than ignorance. What if we cut, and the plant dies? And so, we obfuscate, and wail that we simply don't know where to begin – how fortunate for triage, which provides the ideal point of departure. We could leave it there, content that we've overcome our reservations and done what we can to secure the health of our plants for the next season, but by this time, the secateurs are beginning to feel comfortable in our hands, our confidence is quietly building, and we do after all know a few principles that seem to apply universally, whatever plant we're tending to. We know where and how to place a cut in relation to a bud, we can recognize new wood from old, and we know that cutting stimulates growth, and where that growth is likely to occur. Before long, the issue facing us has changed from how to begin, to knowing when to stop, and the thought begins to suggest itself that it wasn't so much confidence or information that we needed to feel in overcoming our fear of pruning, but joy. Joy in the acquisition of a new skill and new understanding. Joy in the tactile pleasure of using simple tools for straightforward tasks, and in time spent in mindful occupation. Joy in developing intimate acquaintance with the plants that are a part of our home, in the prospect of selectively carving light and space into the garden, of arranging mass and volume in dense pillows and layers after the form of Japanese *niwaki*, or in the airy, transparent style of *sukashi sentei*. Joy in helping to form something that we know will change after we're done, and in having some input into how that change will take place. Joy in collaboration and partnership; and in the recognition of our contribution, given rousing testimony to by the garden's response.

For a gardener inclined to vacillate between a state of quiet meditation and enthusiastic industry, it's always worth asking how vital any operation is to the enjoyment of the garden, and with so much coming at gardeners to tell them what they *should* be doing, it helps to have a pat response with which to buy a little time while we consider the answer. So, the question of whether pruning is necessary, needs to be met, before anything else, with the reply: "For what?" Not for the plant, that's certain – it will grow in its own way quite happily, without undergoing surgery, and there is wonder in such wildness. But if we want to control a plant's size and shape, if we want to increase the yield or the size and quality of its fruit, if we want to eke the maximum degree of flower and glory from our mock orange and ensure that the flowers on the lilac can be enjoyed by creatures shorter than a giraffe then, certainly, a little judicious pruning will be necessary. Pruning involves work, but busyness, of the kind where we expend energy for little or no return? I don't believe so, for the same reasons I don't consider to be "work" the effort expended to hang out with those in whose company I find myself replenished. Pruning elevates our time spent in the garden to a level beyond the roll call of mundane maintenance tasks, and we find ourselves engaged in charming and fruitful conversation with the characters that fill our borders, mingling attentively with all to ensure not just the rejuvenation of old wood, but of our own spirits whilst behaving, to all observers, as a genial and perfect host.

How to DEADHEAD A ROSE

When finally the roses start blooming, we don't want them to stop. But they're only concerned with looking pretty for a while, before turning attention to fruit and seed. Some roses – most climbing forms, for example – are single minded, flowering once a year before getting on with business. Others, with repeat flushes of flower between late spring and late autumn, offer us opportunities to get involved. Removing spent blooms not only keeps the display looking fresh but, by interrupting the process of setting fruit, encourages more flowers.

Secateurs are essential, ideally the long-nosed florist type, particularly useful when space is tight. We communicate with our plants with the blades of our pruning tools – it's only good manners to keep them clean and sharp. You wouldn't want a surgeon coming at you with a rusty old pair of scissors; why would a rose be any less discerning?

By way of introduction, present yourself to the plant. A rose bush is a prickly customer upon first acquaintance, but makes a reliable, loyal, and long-term companion. It doesn't hurt to pay one's respects, or to ask pardon for the impending interference.

A pause. A breath. Fill yourself to the brim with the sweet, perfumed air; musky notes sinking down your legs to your ankles, spicy tones hovering around your waist and high-pitched florals wafting in clouds around your head. It's always good to be conscious about the breath, and in the presence of a rose in bloom, it becomes pure pleasure.

Accustom yourself to handing the rose, placing your hand gently behind a flower or stem you want to work with and gently coaxing it towards you, rather than grasping it with your fingers. The gloves you might think essential for this task most often prove useless; either too thin to provide any protection, or too thick to allow the dexterity required. Getting used to the sensation of thorns (technically, prickles) on flesh inculcates a greater state of intimacy and understanding, but gently does it.

Select a stem with a spent bloom, and either carefully snip the one faded flower out from between fresh flower buds, or remove the whole flower head if all are past their floral prime. In the case of the latter…

…track down the stem from the flower head with your secateurs until you reach the first leaf, and snip just above the node.

Whisper your thanks and remove spent flowers to the compost heap.

Wipe secateur blades with a little rubbing alcohol, apply a drop of oil, and run a sharpening stone along the cutting edges, before putting them away until needed again.

The perfect host

You've decided you want to be involved in what goes on out there in that space around your house, ready to take on the mantle of "gardener" and to practise some intentional intervention in the borders. Whether that's going to be an intensive affair or something lighter of touch, you gradually realize you're throwing a continual party for plants, hovering somewhere between event planner and emcee and, as all the best hosts know, this is a role requiring unstinting attention and watchfulness combined with the ability to project an air of consummate unflappability. To the uninformed spectator it may appear as though you're doing precious little, but you'll find yourself constantly mindful of the contents of your flowerbeds; who's getting along with whom, whose signs of incipient unruliness might benefit from your gentle redirection, and how best to make new introductions to ensure the whole thing goes off with a swing. Naturally, you want your garden to offer a memorable experience, but for all the right reasons. You're there to make sure everyone plays nicely together.

Sometimes I think it would be fun to stand by and watch it all unfold, metaphorically to throw wide the garden gate as an invitation to all comers, though most, if I'm honest, would be already present, and few of those who'd yet to arrive would feel the obligation to enter via such a prescribed route. An unruly free-for-all where each grows as it wants, until checked by a neighbour, at which point some kind of negotiation would take place, accommodations reached to mutual benefit, or something altogether less polite. The closest we get to this here just now is "No Mow May", initiated by British conservation charity Plantlife to encourage homeowners and local authorities to support biodiversity by leaving grass uncut for an entire month, at the end of which time the organization's "Every Flower Counts" survey logs the national stock of wildflowers, plants, and fungi. Three weeks into April, I've yet to rouse the mower from its hibernal slumber and feel disinclined to give the lawn a pre-campaign haircut, having grown attached to the cheerful splashes of white and palest pink from daisies and lady's smock, the latter rarely getting the chance to flower when the grass is cut with a frequency of which the neighbours would approve. There is also more grass than I had realized in my lawn, which I'd long believed to have been almost entirely taken over by ribwort plantain and clover. These still feature in some number but won't assert themselves until early summer.

There are few of us who both enjoy our gardens and view the process of gardening entirely passively and, while the playbook for vegetative succession has long been established in the memory of the landscape, we tend to intervene long before our own domestic parcel of weedy grassland transitions to scrub. We form ideas about the plants that should grow in this space according to need and fashion, rarely giving much thought to the resident populations and careless of the fact that even a garden stripped back to little more than turf and bare earth is less blank canvas than one over which a thin layer of whitewash has been daubed, fading to transparent as it dries and bringing only a short time's respite before the complex underlying picture is again revealed. And into this fertile picture, we make our introductions, at which point the previous residents become reclassified as weeds, and we spend half our time in the garden hauling them out whenever they have the temerity to appear above ground.

Whether we admit it to ourselves or not, what we're really hoping to discover when we visit the garden centre is a showier plant that makes itself quite as at home as the bindweed and speedwell that romps its way through the borders, whilst demonstrating manners decidedly more refined. Because, even at the height of summer, we never have to water the weeds, and yet they continue to flourish in the face of whatever the weather throws at them. I'm uncomfortable with the level of disdain shown for the flowers of the land in favour of plants whose roots, so far from home, may once have yearned for different soil. But I'm as likely to refuse a rose or hydrangea a place in my garden for their Central and Eastern Asian ancestry as I am to let a nominally "native" vine grow away unchecked, winding and throttling its way towards the light. One of the most important roles of the gardener is to locate the delicate balance between plants and place and hold the garden at that precarious tipping point.

Like it or not, discomfort dogs the conversation about the origins of the plants in our gardens. It's both inescapable and instructive that the language we use around the issue – with its categorization of imported plants as "non-native", certain troublesome introductions as "invasive", and a tendency to overlook the problems caused by plants, such as bracken, that are both native and invasive – mirrors that used daily to describe both the movement of peoples and structures of power that have shaped our world and thinking. But a double-standard is waved by a gardening nation that nullifies its own indigenous wildflowers and plants – the commonplace, the weeds – in favour of the more interesting and exotic. More, those who cry "woke" while denying horticulture's imperialist past live in denial, and even the rest of us, while facing off the onslaught, are met with a dilemma, since gardening inescapably involves bringing our own transitory power to bear upon the land from a position of relative ignorance, with a palette of plants gleaned from entirely different climates. When seen in this light, our arrogance is as astonishing as our gumption is impressive, and it's both a wonder and a boast that we manage to grow a thing. Surely the best we can do is to hope to learn, to act as stewards of the land rather than as upstart monarchs over our diminutive domains, and to behave with humility and sensitivity towards the natural world and all it contains, including our fellow gardeners. It behoves us to educate ourselves about the context in which so many of our favourite plants originally evolved – not merely as a horticultural imperative, the better to suit the plant to appropriate growing conditions, but out of respect for the cultures from which they've been acquired, however such acquisitions took place. Within the increasingly politicized "culture wars" debate, institutions such as Royal Botanic Gardens, Kew have been grappling with the colonial origins of their collections; in the case of that particular organization, seeking how best to address this within the remit of its original terms of reference. For ourselves, and notwithstanding the universal value of the system of pig Latin binomial classification first advanced by eighteenth-century Swedish botanist Carl Linnaeus[15], we could do worse than resolve to acquaint ourselves with the local names of the plants in our gardens (*ajisai* for hydrangea in Japanese, *acocoxochitl* for dahlia in the Uto-Aztecan language of Nahuatl), an undertaking accompanied by the delightful notion that someone, somewhere else on the planet, might even now be discovering that another of the common names for the plant British gardeners know as "cleavers" (*Galium aparine*) is "sticky willy".

Such a diverse community of plants, rubbing shoulders with one another just beyond the back door. We tend to talk of its members as though they spontaneously manifest in our gardens, or at least, spring from the ground. But, as we've been discussing, this is the very thing that many of them *don't do*, requiring instead an introduction and suggesting that, at some point, someone had to do the introducing. And if not us, then who? And for what reason? It's one thing to consider how best to ensure all our garden guests get along together, but it's worth spending a little time in considering how they got invited to the party in the first place. In the vanishingly unlikely case that we've been left detailed records of how our gardens have been planted before we inherited their care, we can only surmise from our own behaviour the way their contents arrived.

It's a rare householder who doesn't feel the urge to add to a property's stock of plants, and it has less to do with gardening than with staking a claim (though you could argue that those two activities are more closely related than they might at first seem). We arrive and plant a flag, only the flagstaff grows roots and begins to leaf up before bursting into flower and gifting us with a rose, an apple, or the waxy bloom of magnolia. From the humblest shrub to the most gorgeous of flowering trees, the precise nature of the plant is immaterial; what matters is that something has changed, and that change coincided with our arrival. We have confirmed that the land is fertile and announced, more to the point, that it is ours (or, for the less easily self-deluded, under our care). And maybe that's as far as it goes; but we're gardeners, you and I, and we're not about to stop with a totem.

Adding to the cache of plants already present on the ground, either growing or biding their time lying dormant in the soil's seedbank, are those brought in and planted by builders, landscapers, and previous owners. We're stuck with the first category – the resident plants – or at least doomed to meet at regular intervals during proceedings, but we can exercise more discretion when it comes to the second group. It might seem as though a host's primary responsibility is the welfare of each and every guest but in reality, this is outweighed by their obligation to the smooth running of the function itself, calling for a certain ruthlessness when it comes to those who might already be on the premises before the guest list is even broached. At the point at which we take on responsibility for what, in essence, has been someone else's garden, we have the opportunity to interrogate every plant present to help us decide who is a key member of the household, and who merely a hanger on; who, of the assembled, could have a significant part to play once everything gets going, and who's just going to drag the whole mood down? It doesn't do at this stage to be sentimental or give undue weight to longevity of stay. It mattered not how long the old apple tree had been here when we arrived, how it had been felled by a lightning strike some years previously and refused to die or reduce the abundance of its annual crop; sprawled in ungainly fashion across half the garden, it had to go. And yet the three huge firethorns that had been given licence to grow unchecked across half of the hard-standing at the entrance to the garden were allowed to remain, subject to a severe prune which saw them first pleached, and now clipped into tight green globes floating above sinuous stems, as much a feature spot-lit at night as in the bright light of day. "What can you do for me?" and, "Tell me why you should stay?" are questions to be asked of every legacy plant, assuming you have the requisite steel to take action following an unconvincing answer.

To what remains, we make our own introductions. The best of these come from close friends and family and arrive without ceremony or fanfare, a few sticks in a pot or a soily clump of roots and bedraggled leaves in a bag, the unlikely repositories of memory and identity. For the sake of the stories they hold, and sometimes even for the plants themselves, we take pains to make the garden work around these heirlooms. Then there are the plants we choose with some deliberation, having encountered them in person when trawling other people's gardens for suitable candidates, or via that process of proxy enchantment peculiar to magazines, catalogues, and social media; the gardener putting blind trust in their Holbein while the plant places theirs in blind luck, and both hoping for a happy match. Assuming completion of the necessary homework, and the situation and conditions of one suited to the preferences of the other, there should be every hope and prospect of an outcome sufficiently happy to make even a Lord Chamberlain purr; everything you could wish for as a corollary of informed decisions made by cool heads. But the ranks in our borders are swelled further by those whose introduction comes about through choices made in the heat of the moment, when we've been snagged at our most vulnerable, with hands halfway to pockets and in the grip of that crazy, what-the-hell attitude on which our economy thrives. We are out in force throughout the spring buying plants. It's what we do after a long and dreary winter, especially at the weekends. In the UK, we're lucky enough to be blessed with no fewer than four bank holidays (including the Easter weekend) between March and May, each one a critical date for the garden retail sector. We buy well, if not always too wisely.

When we're ambushed by those hankering for an invitation at the point of sale and caught in the full glare of their seduction, our reputation as the perfect host hangs in the balance. The solution is never to go shopping for plants – never to pass so much as a roadside plant stand – without a clear idea of what the garden needs, and the resolution not to be waylaid. For most of us, this is impractical – at least at first, while we're working out how the whole gardening thing works. During these early stages, we don't quite know what we like, and shopping for plants is part of the learning experience. So what if we behave like magpies with an allowance for a year or two? But there's little longevity in this kind of approach, and the roll call of those initially showy but short-lived specimens grows ever longer, as the confidence we have in our own prowess as gardeners retreats in opposite fashion. If we can hold off berating ourselves over a lack of green fingers, we might realize that we're not so much exercising our choice over the plants we buy, as handing over money for those that somebody wants to sell us, and that these are not remotely the same thing. Sadly, many of us simply believe – falsely – that we're hopeless in the garden.

But in a retail situation, the odds are stacked against us, despite what the market would have us understand. Should we arrive at a garden centre with generous credit limit and an inclination for nothing more specific than "a little colour to liven things up" we will find ourselves happily accommodated, with bedding and plants in the peak of flower racked high at the entrances, at pinch points in the store, and en route to the checkouts. Amongst all the candles, kitchenware, and fleecy clothing, it is the job of garden centres to sell us plants and to keep us coming back for more but, though notable outliers exist, we can hardly expect every outlet to offer us personal planting advice. The exception is the independent plant nursery, invariably staffed by those entirely ill-equipped to keep secret a passion for the plants they grow, and utterly incapable of completing a purchase without first submitting the customer to a full interrogation on the proposed planting site, the better to ensure their precious charges are bound for a suitable new home. In the absence of such horticultural paragons, we have a responsibility – to our bank balance, our notion of ourselves, as well as to our gardens – to ensure we bring to bear upon our shopping that same intention and purpose we have recognized as being fundamental to our gardens. And this means having a plan.

For most of us, such strategies don't spring into being fully formed, and while we might be only a mouse click away from a pleasingly harmonious selection of tulips, the ideal year-round combination of plants to enliven balcony or flowerbed, let alone an entire garden, is rarely available as an off-the-shelf solution. Even those with a flair for how colour, form, and texture can be expressed through leaf, bloom, and bark must pause for thought when it comes to extending the effect – not merely from one side of the garden to the other, but in such a way as to persist from one season to the next. All gardens are sensory, and those that aren't have simply yet to be planted in such a way as to offer stimulation beyond that which could reasonably be expected of a yard, though that purely functional state deserves to be cast in no shade. And so any of us, given the time to think and the mental space to set aside any practical objections, should be able to decide upon the moods and feelings we would like our garden to invoke, if not quite yet the plants we need to make that happen. We can start by matching plant characteristics to the feelings we want to encourage – pale-petalled flowers and luxuriant foliage for calmness and serenity; large, bright blooms and vertical forms for dramatic exclamation; tall and voluminous plants to create privacy and a comforting sense of being enveloped by the garden – creating a structure to guide us when building that community of plants with which we want to surround ourselves, and guarding against the inevitable disappointment of impulse buying. And the expertise to fit the plant to the mood needn't be our own, now that we have criteria of a kind to delight an archivist or cause the index pages of a plant catalogue to flutter in anticipatory thrill. Paired with information on our garden's soil type, moisture level, and available light, we continue to narrow down the profile of those plants best suited to the kind of garden we want to bring into being.

It's a data driven approach and, though it might seem a little cold, it could be considered a kindness to rule out those plants that would have little hope of flourishing or being appreciated within our particular garden setting. Data, after all, is information, and a gardener in possession of that talent for mixing things up peculiar to the perfect host needs chapter and verse on every new introduction, the better to understand how they might behave in company. Of those plants identified as constitutionally suitable for inclusion, the next consideration is temperament – tulips and dahlias showy in short bursts, structural evergreens providing a more softly spoken but sustained contribution to the character of the garden. Some manage to combine more than one aspect – a Californian lilac lending the smoky blue haze of its flowers to the garden for a few weeks in late spring, while making a contribution of deep, glossy foliage throughout the year and, in doing so, straddling the divide between those standout characters that provide the event and incident necessary to make the whole thing go with a swing, and those responsible for the level of background hubbub that provides the assurance of a well-attended gathering worthy of your time. Both personalities are essential to the mix, and it would make no more sense to invite every single loud and brash character in our address books than it would to have a room full of keen listeners, with no-one to regale them with anecdotes.

Interventions are inevitable, such as when the low-level murmur of ground cover swells to a throaty roar, and the creeping bugle over-reaches itself by wandering into the lawn, or the pretty periwinkle, not content to arch and bloom in her allotted space, begins to entertain territorial ambitions. Routinely, it's the ones that seemed most quiet to begin with that pose the biggest threat to harmony, though while the lovely foliage of a variegated ground elder will creep beyond its envelope, a showy alstroemeria or goldenrod won't be far behind. Others, from geraniums to rudbeckias to bamboos that would rather clump than run, seek to expand their spheres of influence in concentric circles, but the best behaved just sit there, often showily, but never quite sufficiently at home to wander from where they've been planted and, though this comes as a relief in terms of flowerbed control, there's part of me that wishes them the confidence to make a nuisance of themselves. Still others refuse to share their space, their strongly antisocial vibes fortified at the roots by chemicals inimical to any who might feel an urge to draw alongside. The allelopathic Mexican marigold might be a stubborn loner, but it's a useful introduction into areas where pernicious weeds prove reluctant to relinquish their hold upon the ground, the soil clinging stubbornly to its own notion of what it wants to grow.

This is a role we take upon ourselves when we decide to garden. Because the garden has its own idea of the plants that go together well, and nature will assert this order at every available opportunity, though her aesthetic may well not always align with our own. Not content to yield this area of responsibility, we take it upon ourselves to provide alternative schemes, and the push and pull of the gardening relationship gains much of its energy from this opposing dynamic. We find ourselves continually on the back foot, working out a rationale as we go, part of the natural world yet setting ourselves up in variance to its order, the best of our schemes an abstraction of what we see in the world around us; pitifully inferior as a simulacrum of wilderness, but unique as an expression of the meaning we make in its presence. Spring flings open the doors to this space in which we have chosen to direct the proceedings, assigning to ourselves the oversight not merely for the conduct of those present, but for the composition of the guest list itself and, as the season passes, our self-image as the perfect host shimmers and slips as we labour to keep up with changing events and conditions; the comfort of those recently introduced, and the reappearance of old and faithful friends that we've forgotten we'd invited. Gatherings that had been restrained, sedate and airy affairs back in winter are now noisy and so well attended they begin to spill out from their original venues to fill the spaces in between, guests from one event beginning to mingle with those from another, and all around the burgeoning of life in burrow, nest, and bud, and every bed and border fattening with foliage and flower. But surely it was vanity to think we ever stood a hope of managing this elemental glee – to suavely flit from plant to plant, to tender "how-d'ye-dos" and keep one step ahead as every day grows longer. And, laughing now at the ridiculousness of such conceit, we surrender to the rolling tide of green, borne onward on a spray of petals and contagious waves of joy towards the dazzling brightness of approaching summer days.

FLOWERS

to nature

Suddenly it's summer, and it seems like I've been moving in slow motion until now. Everything is stretching, thrusting, bursting out, and hurrying on to flower; a not-quite-panicked but still urgent rush to be in bloom and summon bird or bee or breeze to do the desperate deed. Reciprocity reigns; and while there is no courtesy among the plants, understanding lingers in the air between the flower and the bee. Naturally, I want them both, though not yet the consequences of their partnership. But the inevitable can be delayed, and I step in with a *snip*.

You, I've noticed, are not so sentimental; keen to get the flowering done and move on to what's next. I want to pause it all – gold dust, bright light, confetti – to linger among sunshine and petals and pollen. We both know, in time, you'll get your way, but give me just this moment here.

Sun-kissed soporific

Long days and short nights, the temperature discrepancy between the two contracting as spring gives way to summer and the warmth lingers on the evening air. The year is building to its climax, and every centimetre of the garden knows it. You can grow anything now. Pulling the tip of your finger from the soft yielding soil to drop a large, flat seed into the impression left behind, you gently backfill with a flick of the hand, and douse the surface with the fine spray of the watering can. Within a day or two, a strong green shoot thrusts its way upwards, wearing the casing of the seed at a jaunty angle; the speed is breathtaking. Fast forward a few weeks to bright yellow flowers, long, dark green fruit appearing overnight, and your own newly acquired insouciance towards the bountiful production of your zucchini plants; young, finger-length batons destined for stir fries, left on the plant to mature for a day into something suitable for slicing into a ratatouille. Marrows before you know it. If a garden is all about growing, then this space has well and truly settled into its groove, and your own confidence as gardener swells along with the contents of the beds and borders. But is that really all there is to it? Make no mistake, this is a land grab on the part of nature, but it's also a moment of reckoning, an opportunity for you to decide how you want to be both in and with your garden and, having decided, to relax into that relationship. To seek fulfilment through busyness and control, or to inhabit most fully that state of hazy indolence? There is no objectively correct response, but there's a right answer for each and every one of us.

This is the most sociable time of year in the garden when, with windows and doors flung open wide, the division between inside and out becomes little more than a formality, and we look for excuses to welcome company around. In these heady, long anticipated weeks at the high point of the year, a space that bore witness to the private and solitary musings of winter and the quiet but purposeful preparations of spring becomes a venue for fun and games, for spending time with family and friends. I want summer in my garden to be filled with laughter and loud conversation long into the night, the scent of plants kissed by the sun mingling with smells from the kitchen, al fresco meals for days, ever-more adventurous salads, and warm bread dripping with thick, peppery green-gold oil, leisurely morning coffees and, at the end of the working day, long, cool, jewel-coloured cocktails with half a herb patch floating about between clinking cubes of ice. I want clear blue skies and bright sun, air heavy with a hint of breeze and the warmed leaves of verbena and pelargonium offering up their potent greenhouse fug of lemon, rose, and eucalyptus, blended with the astringent musk of tomato leaves I had to wait until adulthood to realize that I loved. I want to feel days stretched out to impossible lengths with time for thought and toil and dreams, and – such luxury – room for companionable interruption, fading with the light into that slightly sozzled, rambling chatter in which erudition is happily exchanged for fellowship, as moths flit between pale flowers and flirt with candlelight.

On the hottest of days, I think I could lie on the ground and while away whole afternoons in hazy half-wakefulness, gazing up into that most excellent canopy through glistening golden oat grass as it sways in counterpoint to hemp-leaved mallow's small pink blooms upon their wiry stems. This pair, together with violet *Verbena bonariensis*, seem to have been designed entirely for the gardener to draw elegant diaphanous veils across the view, coyly hinting at what might be beyond while dancing in a breeze too slight to be felt by the clumsiness of folk. Bees hum from spire to purple spire among salvia, lavender, and toadflax, and mirrorball white alliums bring a chilled-out kind of disco vibe to the twinkle and froth of biscuit-coloured tussock grass. I think I could lose myself in this warm and soporific reverie, but five minutes in full sun reminds me that I'm not built for this lark, and I retreat to a shadier spot from where I can look out upon the dazzling scene.

I have learned to love this time of year – a necessarily accomplishment for those – atypically few of us, it seems – not constitutionally disposed to welcome in the brightness of the sun in its intensity but finding themselves obliged to work day long beneath its glare. A positive advantage during winter and a joy on cool, bright spring and autumn days, a body that runs hot will cause its owner to flag in summer's heat, until the knack of slowing down can be acquired. I don't recall quite how it happened – certainly there was no conscious effort, merely the realization that awaited me one sweltering afternoon at the top of a ladder, perspiration beading my arms and sun cream stinging my eyes as the trimmer's blades swept along the top of an escallonia hedge sending spiced-orange fumes into the air, that I had settled into my summer groove. More than this, I began to notice that the heat now bothered me less than it did those around me whose ability not only to function but to flourish throughout the summer I'd previously regarded with uncomprehending envy. It was something I credited less to purely physical acclimatization – by the time of my hedge-top epiphany a good 40 summers had been and gone without my having made peace with anything much above a barely balmy 20°C (68°F) – than a mental adjustment brought about by necessity. There's a pace to summer gardening that has nothing to do with reducing productivity, and everything to do with eliminating rush and, with little I could do about my own internal thermostat, my mind eventually made the necessary adjustments. Curious as to what had changed, I began to observe a certain economy of movement, a tendency to breathe more slowly and with deeper inhalations, as well as a stoic acceptance of less than optimum conditions. I had, after all, quickly become accustomed to working in the cold and the driving rain, and summer's heat, while being uncomfortable and at times disorienting, at least had the advantage of not turning every flowerbed to mush. On the hottest of days outside, when most of us will still gravitate towards the shade of a tree or the sun-dappled coolness of the woodland edge, a jobbing gardener will sometimes find themself in the company of mad dogs and Englishmen, and out in the midday sun. It helps if you can reach an accommodation with the warmest of weather, as well as the wet and the windy.

As the days pass and the increasing warmth of the sun's glow encourages the gardener finally to emerge from beneath the last of their insulating layers, the strength of its light, especially in those ultraviolet wavelengths unseen by you and me, demands considered response. To T-shirt, shorts, and heavy boots (there's yet to be a lightweight solution to the danger of sticking a fork through your toes) I slap on the same high-factor SPF I've used year-round, though with more frequent applications and slathered over wider swathes of flesh. To protect the eyes, a peaked cap and dark-lensed goggles, a scarf to guard the vulnerable area on the back of the neck, and bandanas tied variously round wrist or head. From the reflection in the greenhouse door, a thoroughly disreputable and dusty-looking character stares back at me; a jumble sale pirate marooned too far inland, gulping gratefully from a dented water bottle.

These are the not altogether ornamental effects of summer upon the gardener but, as we wilt in the heat of day, there's companionability among plants working as hard as we do to protect themselves from ultraviolet radiation, flooding light-sensitive tissue with their own version of sunscreens to limit the damage sustained to their cells, while expelling surplus light energy in the form of heat. Summer can be something of a double-edged sword for the garden – an abundance of light that brings both life and potential destruction, and a level of heat that accelerates chemical reactions, up to the point beyond which, at about 40°C (104°F), the rate of photosynthesis drops. To heat stress can be added that caused by drought, felt first by the most recent arrivals whose shallow roots have yet to tap reserves of water deep beyond the desiccating attentions of the sun, and then by more established plants with every passing rain-free day. Maturity is not without other benefits, and plants that have survived the springtime attentions of sap-sucking pests feasting upon their soft, new growth now benefit from the combined effect of tougher skin and the arrival of beneficial insects, each ladybird or lacewing larva ravenously munching its way through hundreds of aphids a week. In a functioning ecosystem, life and death are finely balanced, and those points in the year that promise most can be marked for the gardener by sudden and surprising loss as much as generous reward.

But we have secured an enduring role for ourselves and, while the irrepressible vigour of weeds gives testimony to the ability of the natural flora of the land to withstand whatever the weather might have in store from one month to the next, all but the hardiest of our purposeful plantings are less robust, their vulnerability requiring our constant attention, and in nothing more so than in the matter of hydration. In the height of summer, it would take a concerted effort to overwater a plant outside, whether planted in the ground or in a free-draining container though, with a mind to the conservation of a precious resource, it would be responsible not to try. It's a commitment to be built into our daily routine, ideally in the cool of the morning or the evening when the pressure of water loss from soil and leaf is reduced and, over a moderate area, it can be almost a meditation undertaken before breakfast, or at the end of the day with watering can in one hand and cold beer in the other, to a soundscape of birdsong, neighbourly chatter, and the radio mumbling, sounds of ball on racquet or bat punctured by gentle applause. There's no escaping the chunk that manual watering takes out of the day, the more so with each additional container or row of veg, to the extent that automatic timers on the end of soak hoses or drip irrigation systems that can deliver water straight to the roots begin to look like an attractive option for more than just the fortnight spent on vacation. The garden sprinkler, as tempting as it might at first appear, is far from an ideal solution for watering anything other than the lawn in summer, when the sheer volume of foliage hinders water falling from above from reaching the soil, while increasing the likelihood of mildew. Besides which, there are less wasteful ways to water and, as global warming encourages ever-higher summer temperatures, the increased likelihood of hosepipe bans will rightly bring an end to the sprinkler's indiscriminate use.

If we justifiably feel a little tied down by the lack of drought resilience displayed by our own additions to the garden, the solution lies largely in our ability to build resilience into the soil by improving its capacity to hold on to both water and nutrients all year round, not least through the application of organic matter. Organic mulch applied earlier in the year has the convenient effect of improving the texture of both heavy soils and light, opening up the structure of the one by encouraging drainage and giving clay particles something to bond to other themselves, while retarding the passage of water and nutrients through the other, giving roots the time and opportunity to benefit from both. Nature takes pains to attend to mulching of her own accord, but the gardener has a frustrating habit of removing her efforts in the name of tidiness, and through weeding. At a time of year when the surface of the soil is hidden beneath the signs of burgeoning life, each patch of bare earth created as a result of our well-intentioned industry presents a vulnerability crying out to be buried once more beneath a lavish layer of compost, the better to prevent the loss of water to the atmosphere while encouraging a rich and varied population of soil biota.

While mulch bestows its benefits from above, being incorporated into the soil through the action of earthworms and other soil organisms dragging it down into the ground rather than through mechanical cultivation, a technique known in Germany and Eastern Europe as *Hügelkultur* can provide irrigation from below. According to this method, raised beds are constructed by arranging decaying wood and other compostable materials into a long mound, before dousing the pile in water and finally covering it in deep layers first of topsoil, then mulch. In their first year, the beds need regular watering and are best planted with crops which, like legumes, are able to fix atmospheric nitrogen in their root systems to counter the tendency of rotting wood to monopolize that essential plant nutrient. But in the second year and beyond, the saturated wood acts as a reservoir for the bed, as well as a self-sustaining source of rich and varied microbial life, creating a productive, low-maintenance growing environment for the garden that requires neither additional irrigation nor feeding.

It's a solution particularly suitable for hungry crops such as zucchini, pumpkins, and other squash – we tend to call plants "hungry" when we want a lot from them, as if somehow the fault is theirs. In truth, when we rely upon plants to fill our bellies, the very least we should be able to do is to stand them a good meal every few days and see that their needs are met. The same goes for those plants we grow for abundance of summer flower; even those – the cosmos, dahlias, and zinnias – from less temperate climates than our own, where their roots have the opportunity to delve deeply for the nourishment to sustain. These Central American daisies might reasonably be assumed to shrug off the most intense attentions of the British sun but, denied a degree of maturity through the combined menace of slugs, snails, and frosty, damp weather at either end of the year, none but the hardiest Mexican fleabane manage to avoid the indignity of being fussed over as a tender annual or, in the case of the dahlia, a perennial whose survival through the winter becomes the subject of an annual lottery – whether to lift the tubers and store them safely in the shed, or leave them in the ground, well mulched, and hope for the best. All too often the plants we are encouraged to love – whose flowers and fruit, pictured in all their summer splendour, were such an enticement to us in the short days of winter and early spring – are now adding significantly to the list of demands upon our day. Surely no one, including the plants in our garden, can much enjoy the reluctant ministrations of the hard-pressed or resentful.

Why, then, do we continue to buy the same selection of seeds every year, with minor variations? Why, when our degree of time poverty has become something about which it's become acceptable to boast, do we persist in setting the garden up as a place of work and busyness, instead of as a venue for observation, involvement, and the community of kinship? It's as true for the allotment and the veg plot as it is for the flowerbed – an endless roster of annual plants that require increasingly more of us; from sowing and germination to potting into containers of increasing size, from being tentatively moved from sheltered conditions to an outside space for "hardening off", and then, finally, planting into the ground or a yet larger container, there to be coaxed, with daily attentions, through the vagaries of summer. As a regime, it has much to recommend it to all those fascinated by the growth and development of plants, the magic that surrounds raising a garden from seed, and with the time to indulge such interests. But it's not the only way to garden, and it's being presented as the default.

Nature has designs upon our beds and borders, never more so than now, the final push of the year, a last chance to claim land and thrust hungry roots deep into parched soil, wresting from the groaning ground sustenance enough for a quick crop of flower, fruit, and seed. More fleet of foot than her bipedal gardening partner, and constantly adapting to changing conditions, her selection of plants is far better suited to this space than most of our own, with a resilience that would laugh in the face of our efforts at devising a nominally "low maintenance" approach. No resident of the garden testifies to this degree of ecological adaptation with more audacity than the bindweed which, just as so many of our treasured introductions begin to show signs of flagging in the heat, winds itself up and around any convenient stem, taunting less resilient specimens by rolling out a fresh, green blanket of heart-shaped leaves across its parched and yellowing neighbours before, as a highlight to the performance, bursting into bloom in a fanfare of pretty white trumpets. And all of this, with no additional watering or feeding, and entirely without regard to our attention.

If nature has designs upon the garden, then summer, in its turn, has designs upon the year, stretching out upon either side, mean temperatures steadily rising by the decade and rain, when it arrives, being delivered in a sudden and muscular deluge rather than the gentle, overnight showers we'd almost certainly prefer. Change is inevitable, but loss is rarely a cause for celebration and the summer skies are quieter than they should be, notwithstanding the birdsong and the evermore gentle hum of bees. While the ability to sit outside, to read or to enjoy a meal without being troubled by clouds of curious insects might seem agreeable, the implications of this latest and increasingly notable absence are worrying. A recent study in the UK that measured squashed insects on car registration plates[16] suggested a fall of up to 60 per cent when compared with data from 2004, with a similar survey of windshields in Denmark[17] five years earlier pointing to an even greater decline of up to 80 per cent. With each study taking pains to point out the essential role in the planet's ecosystems performed by invertebrate life, we *should* be alarmed; 60 years after the prescience of American writer, scientist, and ecologist Rachel Carson's *Silent Spring* (1962), agricultural practice and government policy at both national and local levels pay little more than lip service to measures that would reduce this critical trend; legislation has proved toothless, harmful pesticides are still racked high on supermarket shelves, and the peat extraction lobby persists in behaving as if there's a legitimate debate to be had around the destruction of unique and irreplaceable habitat. Our gardens should be brimming and buzzing with life at this time of year; there's nothing calming about this present level of quietude.

There's a crumb of comfort to be taken in the ability of the garden to adapt to the combined threats of a climate in crisis and our own domestic and global behaviours, including patterns of trade which see the widespread distribution of plant pathogens – the fungus responsible for ash dieback; the bacteria *Xylella fastidiosa* laying waste to olive groves in Europe and stone fruit orchards in the US – but the buffering capacity of these individual semi-wild pockets of green around our cities, towns, and villages is finite. An order of change is coming to our countryside of a magnitude far greater than the loss attributed to Dutch Elm Disease that was such a feature of my own childhood in the 1970s, and there's slim likelihood of our being able to garden as our grandparents did, with the same plants and attitudes. While enlightened farmers embrace a regenerative model of agriculture that returns the health of the soil and the preservation of biodiversity to their rightful place at the centre of any land-based enterprise, there's an urgent need for gardeners to stop seeing our activity as a pleasant and optional pastime and begin seriously to acknowledge our role as stewards of the land over which we currently exert some influence. Though the percentage of the land under agricultural management dwarfs that laid down to gardens, collectively we can make a difference that extends beyond our immediate domestic sphere, creating uninterrupted networks of biodiversity while connecting gardens, streets, neighbourhoods, towns, and cities to the countryside beyond, and helping to build the resilience back into our landscape one busy, buzzing, flower-filled garden at a time. It shouldn't be a hard sell, but our evident desire to see the natural world subjugated and placed under our control sits at odds with a truth increasingly hard to deny; that our own needs and the needs of the environment are one and the same.

In spring, surrounded by the vibrant energy of new life thrusting from the ground, it's easy to be full of hope. By summer, stress is beginning to show in the arrival of curious signs to mark the passing of the year, and the lack of familiar ones – the thinning canopy of the ash tree in whose dappled shade we've grown used to whiling away a lunch hour, the yellowing of the leaves on a hitherto robust and reliable shrub as it struggles with a combination of heat and drought; the initially welcome and then, suddenly, frightening absence of an entire cadre of creatures. My own path ahead seems clear: to relax my hold upon any notion of how a well-tended garden should look, to tolerate the weeds whose mere presence I've been taught to consider an affront, but in whose flowers the pollinating insects I want to encourage find such obvious delight. To concentrate my efforts on building resilience back into this space from which so much has been demanded by successive tenants, with scant recompense; to celebrate its natural flora and, where making introductions, to choose plants that combine hardiness, drought tolerance, and generosity of floral display. And recognizing how I garden, to make these choices perennial, or at least, reliably hardy annuals with a tendency to self-seed. To look to cultivated relatives of those wild plants that do so well – morning glory for bindweed, asters and goldenrod for ragwort, so many ornamental members of the nettle family for their stinging relative, though at least a small area will always have its place here, for the butterflies, and the pesto.

The garden is a riot now. All pretence of decorum abandoned, and the greater part of the lawn surrendered to my natural accomplice. At some point I'll find my way to the shed and liberate the mower, cutting a single strip through the long grass that's played host for the past few weeks to daisies and dandelions, lady's smock and – new this year, I'm sure – the tiny white petals of thyme-leaved speedwell with their purple eyelashes. Buttercup and ribwort plantain are arriving for their shift and, as I gaze out across the rich green sward, remembering the parched cropped turf of summers past, a hoverfly descends in front of my shoulder, and peers at what I'm writing. Everyone's a critic. The penstemon, certainly, is not my biggest fan just now, surprisingly resentful of my decision to restrict watering to vegetable beds and containers, with varying results, and I feel a momentary pang of guilt for my willingness to withdraw my fussy attentions and surrender a plant to its destiny. It will thrive, or it won't, and its fate will be my instruction. I've come to realize there's a watchful kind of effort that I'm prepared to lavish upon the garden, and a frantic sort of activity I'm loath longer to extend. The time will come for wholesale taking stock, for addressing the successes and failures of my own contributions to this space but, in the drowsy heat of this summer afternoon's sun, I'm content simply to survey the scene through half-closed eyes and puzzle out what the bindweed has to teach me, to a soundtrack of the birds and bees.

Times tables

It may be the heat, the conspiratorial murmuring of the bees in the lavender or the brazenness of the floral display – flowers are, after all, nothing if not the rude bits of plants – but everything in the garden seems to be at it just now. In truth, it began with the first warm days of spring – out here, it's all been about the birds and the bees though, while the former began the carry-on back in March, the latter will wait until the dog days of summer before attempting anything fruity. (You can't deny a male honeybee his reticence, given he'll get to mate just the once, in midair, after which unrepeatable coupling his penis will explode and he'll fall to the ground, stone dead. Still. What a way to go.) And ever since the first demure snowdrop archly showed a little style back in February, plants have been single-minded in their pursuit of the most basic instincts, to the point where the fuggy summer air is thick both with clouds of pollen and with insects ferrying communications of amorous intent from one bloom to another. Summer lovin's in full swing, multiplication the order of the day.

From where we stand, plants and life are two concepts inextricably intertwined. More than just a metaphor, the currency of flora resides in something greater than its value in providing us with parallels to our own existence, a dance of life and birth and death played out year after year, three-score-and-ten rehearsals for our own. The importance of a landscape full of plants rests not merely in its ability to fill our souls with inspiration and our hearts with song, but to fill our bellies full of food. It's a vital energy that calls to us with a heartiness in inverse proportion to our own, perhaps explaining why the draw of the garden increases upon us as we age; that only the wisest of us in our youth when typically we hold life less as a precious gift than an inalienable right, may appreciate its attractions. Quite what a plant makes of life will doubtless remain a mystery to you and me, though we could hazard a guess in the joint direction of living – or existence – and legacy. It's hard to tell how much a plant enjoys the former, but its commitment to the latter is witnessed in the effort it expends to reproduce and, while spring steals top billing as the season of new life, summer is where conception mostly happens, out here among the petals.

Inescapably messy and often exhausting, the business of sex seems like the logical starting point when it comes to any individual's personal quest for legacy. But for many plants that can't be bothered with the fuss, there's a lazy workaround; a way for them to increase their hold upon the planet without so much as breaking a sweat. For a human to enter a vegetative state would have dire implications, but it's second nature for a plant, so why not attempt the process of multiplication through methods at once both familiar and untaxing? A little surreptitious stretching, a modicum of flex and extension, a gentle pushing against the boundaries – the creeping expansion of a plant's territory often goes entirely unnoticed until new individuals reveal themselves, standing tall and proud in locations far away from the original, having made their way through the underground creeping of adventurous root systems or similarly intrepid stems, edging along the surface of the soil. A beautiful chocolate vine I once tended caused no end of perplexity by its refusal to flower despite being situated in the ideal spot until, scrabbling about in thick layers of mulch, I realized it was living its best life by sending out long shoots at ground level, and gradually laying claim to the back of the border. Feeling a little mean, I panicked it into flowering by cutting it off from its colonial acquisitions, having correctly assumed that, should the easy way to procreate prove ineffective, it would consider putting a little effort – and flower – into the enterprise. In a very real way, I did the plant a favour since, as a long-term survival strategy, vegetative propagation is not only work-shy, but unwise, with every iteration of the plant a clone of the original.

This is high-stakes growth, a reckless expedience for those with a devil-may-care attitude to risk. Without the insurance of genetic variability to provide a buffer against attack from pests and disease, there's a very real chance of the kind of population collapse seen with the so-called "English" elm (*Ulmus procera*), likely a clone introduced by the Romans from Italy via Spain that, despite producing pollen, rarely sets seed. Its preference for reproducing through vegetative means robbed this elm of the diversity that could have provided a more effective response to Dutch Elm Disease, which had such significant consequences during the epidemic of the 1970s in Europe and America. Far from blaming the plants, we should look to our own historical culpability, and once again to that artifact of empire that has allowed powers of conquest and exploration to exploit and relocate whilst giving little thought to consequence – first in this context with sterile elms transported about the Roman Empire for their value to viniculture, then with sterile bananas.

The banana, thought to have been domesticated around 7,000 years ago, was brought from Southeast Asia, via Africa, to South America and the Caribbean, where it became among the most traded commodities in the nineteenth century and remains so to this day. But edible varieties of the fruit – almost all of which are hybrids between two wild species *Musa acuminata* and *Musa balbisiana* (also known as plantain) – being exclusively seedless, banana production is entirely dependent upon vegetative propagation. It's a system which, when all is well, proves remarkably convenient for commerce, but when disease strikes, less so. Panama disease (also known as fusarium wilt), brought devastation to the industry in the 1950s when the dominant variety in production was ravaged by the soil-inhabiting fungus *Fusarium oxysporum*. After huge loss to plantations based exclusively around the 'Gros Michel' banana, the more resistant 'Cavendish' was developed, quickly taking the place of its more susceptible forebear. But Cavendish bananas are themselves now under threat from a newly evolved strain of the *Fusarium* fungus, reminding us once again that, since nature never sleeps, an over-reliance on cloned plants is hardly a sound foundation on which to build a trade.

Of those plants plundered rudely from one ecosystem to be thoughtlessly introduced into another half a world away, one of the most successful and to date least vulnerable must be Japanese knotweed, a single female clone of which was introduced to the UK by Victorian plant hunters who admired its handsome, heart-shaped foliage and sturdy red stems. Soon escaping from the garden, and with no male specimen with which to reproduce, it was thought that the major threat posed to landscapes both urban and rural by this famously aggressive plant came via the vigour of its root system and its legendary (though arguably apocryphal) ability to break through concrete barriers to colonize wide areas. But recent research has revealed that hybridization has been occurring between closely-related species of Japanese knotweed, giant knotweed, and their equally rampant cousin, Russian vine, suggesting that, when it comes to legacy, nature is more concerned about viability than the very human preoccupation with purity.

Whether it occurs between closely-related species or across different genera, hybridization is as genetically adventurous for a plant as for any organism with the ability to reproduce in this way – a process that sets against the prize of partial heredity the veiling or loss of certain characteristics from either parent, with the additional benefit that viable offspring are likely to be marked by that peculiar hybrid vigour that makes them a force to be reckoned with in their own particular environment. More compatible matches involve less of a gamble and, while they may be more costly than taking the vegetative route, at least provide the security of genetic variability. In a game where survival is the prize and the adaptability of pathogens is assumed, a moving target is much harder to hit. What's wanted is a way to merge, in diverse combinations, the essence of one parent with that of the other and send it out into the world. What's wanted is a seed; a tiny parcel, wrapping up the best of mum and dad, the totality of their jointly-inherited wisdom encoded into a strand of DNA and packed up with sufficient food to last through those tricky infant stages, at least until the nascent plant has the wherewithal to cater for itself. If seed, then, is the goal, we must have fruit, and before that, flowers. (Conifers and similarly non-flowering seed-bearing plants, such as cycads and the ginkgo, go about things in a slightly different way, eschewing the fruit course and, in doing so, garnering for themselves the scientific description gymnosperm, or "naked seed". But the theory's much the same, though the sexual reproduction of spore-bearing plants, such mosses and ferns, is a truly Byzantine affair.)

Somewhere between flower and fruit, we must have pollination, and here we find ourselves in territory at once familiar and entirely alien. The maleness of the plant is in its pollen, each grain containing three cells: two haploid gametes, each with half a set of unpaired chromosomes, and one extra cell seemingly along for the ride. Every guy needs a wingman. Peer into the heart of a flower and typically you'll find two types of structure: the central pistil containing the female reproductive parts and, surrounding this, the male stamens each comprising a long filament with a pollen-producing pad – the anther – at the top. Even within the biggest blossom, there's not a great distance to travel between the two, but the process of pollination involves all manner of safe-guards, protocols, and chaperones. Meanwhile, playing the part of the flowering structures on conifers are cones; male pollen-bearing cones release their issue in vast clouds to the air, relying on wind to act as an essential third party in the courtship and, ultimately, the fertilization of the female cones. The bits and pieces may appear entirely different, but the goal in each case is the bringing together of the male sperm and the female egg, and the union of the genetic information encoded into each.

There's nothing remotely seemly about wind pollination. Being neither the most subtle nor energy efficient way to go about fertilization, it relies upon the ability of the male to release pollen in vast, hay-fever inducing quantities, hoping some of it will find its way to the necessary portions of a receptive female after a manner in which we can only be grateful that no mammal, least of all one of our own species, has ever had cause to emulate. As a method, its adherents are limited not only to conifers, but include grasses, cereal crops, and many flowering trees, all of which may safely renounce ostentation in their floral parts, having a need to attract nothing more than a passing breeze to get the deed done. The female flowers of the hazel are a case in point – revealed to the careful observer by the tiny red styles protruding into space from a grain-size bud and dwarfed in size by the pendulous catkins of the male. The hazel, which carries both male and female flowers upon the same tree, makes at least one concession to strategy in ensuring that it can only be fertilized by pollen from a different individual, thus preserving a healthy degree of genetic variation amongst its issue. Other plants may guard against self-pollination by restricting individuals' expression to either male or female flowers, but not both. Such dioecy comes at the cost of half the population being unable to bear fruit; if winter berries are what you're after, it's advisable to check that any holly tree introduced into the garden is a female (something confounded by some cultivar names – 'Silver King' and 'Silver Queen', for example – having apparently been applied by a fan of the British Christmas panto with its tradition of cross dressing). The reasoning behind this caution is widely, but not universally accepted in the plant kingdom, and some of its members – among them certain legumes, sunflowers, and orchids – weigh the risk of genetic conformity against the benefits of the environmental adaptation of their particular phenotype, either actively preferring self-pollination or falling back on it should cross-pollination with another individual fail to occur.

Historically, in matters of courtship and love, a go-between has often proved necessary. For the most part, plants deal with the restrictions on their mobility with elan, never too proud to seek assistance when the occasion absolutely calls for communion of an intimate nature. But in situations where wind pollination has proved too vague and indirect a method of transfer – in areas of great species diversity, or where compatible individuals are less closely situated, for example – the assistance of animals is sought. This, perhaps more than any other single factor, underlies how our gardens appear, given that it bears directly upon the appearance of flowers and the consequent incorporation of the plants that carry them into our beds and borders. But given that, through a typically anthropocentric lens, we tend to regard the whole scope and variety of floral invention as little more than a concession to human taste and whim, the realization of nature's supreme indifference to our opinion can come as something of a surprise. Though the most dedicated or obsessed of us may, with paintbrush or finger, act as messenger from one plant to another, the entire display is put on not for our benefit, but for that of the small, crawling and buzzing creatures, together with the occasional bird, whose assistance is so dearly desired.

And what manner of inducement is arranged, the better to encourage visitations by the curious, the hungry, and the downright opportunistic, in the hope of incidental pollination? Insects and birds are as susceptible as you or me to matters of form, colour, scent, and design – even more so, some plants luring potential pollinators with petal markings visible only in the ultraviolet spectrum, though the tiger-stripe landing strip on the horizontal petals (or falls) of the iris, give us a clue as to what we might be missing. Plants have often co-evolved with a particular pollinator, their shape and features frequently tailored to the needs of these creatures; the larger, lower petals of lipped flowers such as salvia and nettles providing a platform on which butterflies can land to feed, the flat-topped structures of verbena and achillea performing a similar function, while long, tubular-shaped blooms like honeysuckle and penstemons suiting the long proboscis of the bee or the beak of the hummingbird. Despite an excellent sense of smell, butterflies are thought to rely more heavily on visual signals than the scent of flowers when searching for food, while moths are particularly attracted to night-scented blooms. Attraction, as everyone discovers sooner or later, is key to survival, and in a benign sense (other than that relatively small group of carnivorous plants), a plant pollinated by anything other than the wind finds itself desperate for bodies. Thankfully, many creatures are happy to lend themselves out in return for dinner, in the form of protein-rich pollen, and a long draught of sugar-laden nectar, an energy drink for the workers that must delve into the heart of the flower to sup it from the nectaries deep within. Together with the internal morphology of the flower, the lure of feeding – and, in the case of bees and wasps, of collecting both pollen and nectar for their young – creates a situation in which a visitor is highly unlikely to be able to go about their daily business without being covered in pollen. While the smooth-bodied wasp still manages to maintain a degree of crisp, sartorial elegance on its forays from flower to nest and back again, the hairy outfit of the bumblebee appears to have evolved expressly to maximize its potential for pollen portage, incidental to that expressly collected into the baskets on the hind legs of the female.

Each plant needs more from their pollinator than a basic courier service; a full pick up and drop off is required at each stop if the diversity of the gene pool is to be maintained and the business of cross-fertilization transacted to the satisfaction of all. And it's not enough simply to leave the package in the lobby – the donor pollen, collected by the bearer in the process of brushing past the anthers on a previous stop, needs to make its way to the stigma of the receiving bloom, though the positioning of the female organs at the very centre of the flower makes an unsuccessful delivery decidedly improbable, even for the most fumbling participant. With the issue of one parent resting upon a part of the other, the moment arises for the third cell inside the pollen grain to spring into action, which it does with no small degree of excitement in response to the chemical ministrations of its new environment. A pollen tube begins to form, developing in a shameless, schoolboy-giggle-inducing fashion, extending all the way from the stigma, through the stalk-like style, to the ovary, and creating a pathway along which the two haploid male cells make their final journey. One fuses with the female gamete, or egg cell, to form the diploid zygote, a structure with a full set of parental chromosomes that becomes the embryo. But the remaining sperm cell plays a role just as essential if more transitory, combining with a secondary nucleus within the ovule and, in so doing, giving rise to the nutritious endosperm that will support the plantlet through the perilous process of germination.

Mission accomplished, and any hope of countering an assertion that holds the flower to be intrinsically, obsessively, and exclusively concerned with sex is dashed by the sight of petals cast disdainfully aside. Having no further need for these gaudy rags they are abandoned, either to the wind or to gravity, though the garden receives them gratefully, breaking them down into their constituent parts as a contribution to the soil's ongoing health. In this way, fertilization brings the curtain down upon the flower; though for the fruit and the seed, it's just the beginning, the development of both belonging to the next act of the plant's existence.

This is how things are meant to progress and given the opportunity, they'll do so with impressive efficiency. But, in spite of the testimony it would give to our garden's fecundity, a border filled with seedheads is not a look much prized by the gardener, and particularly so in summer when we want to be surrounded by a profusion of floral exuberance in colour and form and scent, wave upon wave of blossom stretching out in defiance of the inevitable turning of the year, a denial of the passage of time. If there's to be a hope of hitting the pause button at the very moment the garden reaches peak floof we must once more intervene. We must be, for our plants, the agents of delayed gratification, interrupting their progress from flower to fruit by removing "flowered" stems at the first sign of fertilization – the dropping of the petals, the first swellings at the top of the stem – and, in so doing, encouraging further opportunities for pollination by the production of more flowers. Wandering along the path with secateurs and trug on a summer's evening to "deadhead" those blooms with the temerity to follow their natural inclinations, we can do this, it seems, indefinitely – in reality, the plants eventually tire, energy reserves dwindle and the rate of flowering slows as the season progresses, at which point it's a kindness to allow nature to take its course. If it feels as though we're taking advantage of the plants (we are), we can console ourselves with all the extra sustenance we're helping to provide for the pollinators.

Eventually, and despite our interruptions, the garden will go to seed, and all but the latest flowering introductions, oblivious to the implications of approaching frost, will give themselves leave to look strained and ragged, their energy no longer directed towards root and stem and leaf, but into the production of subsequent generations. But we're not there yet and, while we have long evenings, blousy blooms, and the company of bees, we will snip away at what we have planted, while the weeds look on in wry amusement at our attempts to hold back the relentless progress of the year.

How to HEAR A BIRD

I heard a blackbird today. Gold-ringed eye on soot-grey ground, he kept me company as I weeded, scrabbling nosily at the foot of hedges and pulling worms from freshly turned soil. Eloquent in silence, neither chirrup nor trill escaped him all day long and still he lectured me while my back was turned to the work, until the flutter of wing-beaten air proclaimed the day's instruction at an end. The robin, with no such monasticism, fixed me straightforwardly with both eyes and repeatedly announced his displeasure at being second in the queue.

There's always something to listen to in the garden; an aural collage which we can receive in all its fullness, or as individual, isolated streams, according to our inclination. To focus in on any one sound is not dissimilar to tuning an analogue radio, as we direct our receiving equipment to filter out every extraneous signal, or successively silencing each channel of a multi-track recording until left with just the one.

A pause. A breath.

Excepting the middle of the night (and sometimes even then) or the apocalypse (unlikely) there will be birds, though they're more vocal when breeding or defending territory in spring and summer. Today's task is to isolate one voice from the general clamour and give it a few minutes of dedicated attention.

Picture the garden soundscape like a musical score, each instrument with its own stave and clef, which you can gently touch to temporarily mute:

…first the traffic; the aircraft overhead, the roar of the road, the beeping of reversing trucks…

…then the neighbours; the chatter, laughter, and crying of children, the bouncing of trampoline springs, the power tool whine and putter and all din that denotes a life lived in proximity to others…

…then the wind; the susurration of the breeze whispering through leaves and branches and its deeper, more intimate rumble as it flows and eddies around your outer ears…

…then each bird in turn *muted* until, chit-chat and jabber aside, you find the one voice that seems to hold a message, even if not necessarily one specifically for you.

Stay a while inside this sound, long enough to learn its moods and its inflections, the cadence of its song.

But remember that, like the blackbird, a companion need say nothing to be worthy of a hearing. It's all about our presence, attention, and the practice of actively listening.

A constant reminder

Weeds are the plants that would be where our gardens are if it wasn't for us. But the spin is all; at some point, with a kind of settler mindset, posterity did a number on the humble wildflower and transformed its status from rightful occupier to reviled outcast. It's worth pondering just how and when we learnt to identify those plants that we now class unthinkingly as "weeds" – what stories were we told, and from what age? Were we provided with a few reliable signifiers for the weediness of a plant – green and yellow (mostly), showing a marked disinclination to grow in neat, straight lines – or was the label applied by those who informed our thinking as we encountered each new specimen (dandelion, buttercup, groundsel, thistle) over the course of our daily explorations?

The presence of weeds in our beds and borders is a constant reminder of the landscape beyond the garden wall; missives from nature against our continued refusal to see the blindingly obvious. Only think of all the features common to weeds that would be highly desirable as characteristics of the plants we habitually bring back from the garden centre: strong growth, tenacity, longevity, drought-resistance, free-flowering, self-seeding… the list is long, yet we disdain the ordinary in favour of more rarified specimens. If we're to begin to understand the ground on which we choose to garden, we have some unlearning to do.

It's worth spending a moment or two in acknowledgement of the dual identity of these plants; that outside the managed environment of the garden and its agrarian relative the field, the very same plants we refer to as weeds are seen as "wildflowers". You can't have a weed on a hill or in a hedgerow, where's it's simply not possible to begrudge the presence of pimpernel or to harry herb robert – they are there by right. But, buried in the small print of the contract we imagine that we have signed with the natural world, there must be something that insists upon these rights being laid aside in situations where to exercise them would restrict our own liberty to cultivate the land for food or ornament, or even to enjoy the dubious pleasures of pristine asphalt and concrete. In such cases, the epithet "weed" is applied, towing behind it like Marley's chains an inventory of disparaging labels. Unwanted. Cast out. Troublesome. Undesirable. And the clincher, for irony: invasive. How has this come about?

Weeds only appear when there's a difference of opinion over land use. Nature knows what she wants to grow and where, but agriculture requires a change of plan, since ways of feeding ourselves more efficient than foraging have come to mean monoculture. The presence of anything other than the crop species in the field leads, through competition for resources, to a reduction in yield, whilst raising the spectre of contamination and its accompanying implications for taste, toxicity and, by extension, value. Weeds are an expensive problem for the arable farmer; for the livestock farmer, the contamination of feed with noxious weeds is likely to cause more of an issue than animals on well-managed pasture grazing upon unpalatable, noxious plants. It's not hard to see how the classification of plants into the broadest of categories – desirable or otherwise – has become closely bound up with the question of how to feed ourselves. How to charm ourselves, how to bring relief, sanctuary, delight, and inspiration, might seem a matter at once less consequential and more lofty, but the issue of land use is of no less relevance in the garden, and the categories identical; though while the list of outcasts features the same set of characters as before, the desirables are different in a domestic setting where ornament reigns over calorific value, and taste trumps tasty. Here, we declare that we despise weeds for their ordinariness, but we need to be honest about what's really going on – we're just jealous of how well they grow, and perhaps slightly resentful that we didn't plant them.

But we tend not to give the matter much thought, seeming to know by instinct which plants are meant to be there, and which are not. When we come into the world, we take those things that have preceded us into being for granted; the garden, the lawn, the flowerbeds; the flowers, which are supposed to be there, and the weeds, which are not. (There's a tendency to refer to all plants that we think are "supposed" to be there as "flowers", whether they're in flower, or otherwise. It's a dichotomy that embeds a misnomer, since weeds flower just as well, if not more efficiently, than plants grown for ornament or for food, but one that somehow still manages to make it into print.) And that's just *how it is*. And it *might be*, according to our understanding of the tiny portion of the world in which we've spent our existence. But from a less cossetted perspective, that's just how it *isn't*. And if there's a chance that our parents, or our parents' parents were mistaken, or misled – or were farmers who brought their aversion to interlopers within the crop back from the fields and into a domestic setting – if there's a possibility that we're losing out on something by consigning a whole category of plants to the compost heap (metaphorically or otherwise), we'd do well to decide for ourselves whether or not such attitudes bear scrutiny. And to do that with any degree of success, we need to work out just what it is that turns a plant from a wildflower into a weed.

It's safe to say, that without us, there would be no weeds, and everything currently designated as weedy would be a wildflower. But in the garden, we are the ones who get to say what stays and what goes, what dies and what grows, even if this involves an outrageous stacking of the odds in favour of our less robust introductions. For all we love them, it bears repeating that a garden is a truly preposterous conceit. We take a parcel of land, denude it entirely, and replace all its associated flora with plants of our own; having the cheek to demur when the land reasserts its preference. Look at your garden now; there's barely a plant, barring the weeds, that belongs to the land. Even the Great British lawn – which we might like to think of as the kind of low level, pre-scrubby grassland or sheep-nibbled pasture that would be on the spot had someone not plonked a row of houses down – is entirely a construction, populated not with meadow species but a mix of rough rye grass and more refined species in ratios according to soil conditions and intended use – croquet lawn, football pitch, or green baize carpet rarely trod. Gardening involves the attempt to force a symbiosis between the soil and a motley population of plants where none rightfully exists; which makes it all the more wondrous when it works, even if, much to our chagrin, it works better and more fruitfully with weeds. And so, according to this understanding of gardening, we can trot out with some hope of being right that appalling definition that would have us smugly announce, "a weed is just a plant in the wrong place". Because now, to the chorus of heartfelt objections – "but it is in the right place" and, "who says it's the *wrong place*?" – we can announce, respectively and decisively, "not in my garden, it isn't," and, "I do," since what we say goes. Or so we'd like to think. Gardening, as well as being preposterous, involves a degree of self-deception, our Pecksniffian bubbles burst with every dandelion that rears its golden head uninvited.

This is the phony way I spoke of it in the introduction. Riding to the aid of our delusions come the manufacturers of garden herbicides, preaching a gospel of purity and freedom from weedy transgression, adopting as a sound business strategy a two-pronged attack: first, demonizing the indigenous section of the garden population, and then selling to the concerned horticulturalist the means of its eradication. It's such a pervasive position that if we're ever truly to garden in sympathy with nature, we need to have ready answers to those insidious arguments that form its defence. It states *weeds are unsightly*; but beauty is in the eye of the beholder, and pollinators love a weed. *Weeds are weedy*; well, duh. But weedy means resilient, stubborn, and strong, and who doesn't want that in their plants? *Weeds will compete with your plants for resources*; this being true, we need to make peace with the challenges of growing recklessly polite plants, choose vigorous varieties more suited to the prevailing conditions, or embrace a style of planting that incorporates weeds into the heart of the design, an idea espoused by the garden designer

Jack Wallington in *Wild About Weeds* (2019)[18]. *Weeds are often poisonous*; set against the toxicity of snowdrops, daffodils, hellebores, aconites, and clematis, this is probably no truer for wildflowers than it is for our favourite cultivated varieties.

Weeds harbour pests and disease that can transfer to "your plants". This is a point that must be conceded, but while there's no denying that common groundsel, for example, acts as a host to fungi that cause rusts and mildews to develop on plant leaves, as well as the rotting of roots on peas and squashes, it's not a difficult weed to control in a garden setting, and particularly if pulled up or hoed down before it has a chance to set seed. But it's a point that needs to be received as an encouragement to get to know the plants of our soil and the biodiversity they support, rather than being a justification for chemicals whose use will have an irreversible impact on the local ecology.

As an objection, it needs to be set against the many advantages of tolerating weeds in our gardens, not least through their benefit to wildlife and the consequent role their presence plays in increasing a biodiversity presently fending off the existential threat posed by a combination of habitat loss and fragmentation, indiscriminate pesticide use, and the climate crisis. That same scrappy groundsel censured for the refuge it offers to certain fungi also plays host to several pollinating insects including the cinnabar moth, but if the charms of this glamorously-arrayed visitor prove insufficient to outweigh groundsel's disadvantages, we should find it harder to deny the stinging nettle space. At the height of the growing season, this perennial nettle spreads quickly by seed and at the root, though not at a great depth, making it one of the less tricky weeds for a well-gloved gardener to control. But against its vigour and the irritant hairs covering foliage and stems should be weighed the beauty of the company it keeps – the many moths and butterflies including peacock, comma, and small tortoiseshell that make a nursery of its leaves. If this isn't enough, we can remind ourselves of the nettle's many uses to us as a culinary and medicinal herb – one that grows quickly and unfussily and provides handfuls of bright, young foliage packed with vitamins and other beneficial compounds to relieve hay fever and inflammation, to be taken as a refreshing tea, cooked as a substitute for spinach, or chopped finely and processed into pesto. I find myself wondering if many of the plants whose presence we most often bemoan – nettles, cleavers, ground elder, Japanese knotweed, so many not only edible but nutritious – would present less of a problem to our imagination if we were to make more frequent use of them in the kitchen. After all, what's a globe artichoke, if not a thistle with ideas above its station, its benefit to wildlife impossible to deny by any tempted (for reasons of aesthetics or laziness) to let the vegetable flower and go to seed, the electric violet blue of its petals throbbing with insects on a warm summer's day, the same flowers frequented by birds in winter, drawn by the promise of seeds and their downy parachute silk for the feathering of nests. Meanwhile bull thistles and knapweeds, beloved of bees and burnet moths, bring us an eyeful of delight and, you'd be forgiven for thinking, nothing but a mouthful of trouble; but you can eat most all of a thistle, while knapweed's petals – little blue tongues of flame – will add a subtle crunch and the barest suggestion of pepper to a salad.

When it comes to feeding ourselves, we can buy the vegetables we eat or, even better, grow them ourselves. We can send off to small, organic, independent suppliers for seeds of rocket, spinach, mizuna, and lamb's lettuce, delighting in both in the variety of greens we can nurture in a small space and the wealth of flavour available to those willing to eschew the flaccid blandness of supermarket offerings in favour of putting in a very little effort. But the demonization of weeds results in us overlooking a ready supply of flavour and nutrition only a few feet beyond our doorsteps and, with food poverty and the security of its supply being matters of increasing concern, it would seem a dereliction of duty to ignore such a ready resource, to fail to incorporate its use and care into our lives and gardens, or to think of ways in which we can sympathetically harness its benefits. We have shelves buckling from the weight of field guides and books on foraging food from the wild (*take only what you need*); we can avail ourselves of autumnal mushroom walks and workshops to reacquaint us with once familiar plants, yet there's little sign yet of any of us chowing down on platefuls of garden weeds. It might not offer a complete solution – may even find itself politically weaponized as a convenient but empty answer to food banks and value label ranges – but it's a thing to ponder: we could all eat more weeds, reducing our grocery spend while improving our nutrition, or we can spray them with chemicals and throw them in the bin, while continuing our learned patterns of consumption. It doesn't seem like a difficult choice.

There's poetry and beauty here among the commonplace, food for the soul as well as the stomach, from the leonine gnashing of a dandelion leaf to the majestic white bells of the bindweed and, while I'm never not annoyed by the carpet of bittercress that springs up on freshly cleared ground the moment my back is turned, or its habit of spitting seeds straight into my face, I'm continually enchanted by the rhythm of its rounded leaflets, and impressed at its importunity. It tastes good, too, for a tiny cabbage – earthy, cressy, calling for English mustard (also the product of the cabbage family) slathered over terrible plastic ham in thin, white bread, in which it would undoubtedly be the nutritional star. And if creeping buttercup confuses me with its leaves, hiding amid geums and geraniums (it's the white splashes on the leaves that give it away), that's more testimony to the charm of its foliage than a sign of its cunning. It was here first, after all. But however handsome our own flora might be, we remain unimpressed, while importing other people's weeds.

British gardeners embrace North American goldenrod but reject rosebay willowherb – the one yellow, the other pink, both summer flowering and able to rise head and shoulders above the throng before the blooms explode into a mass of fluffy seed. Both are pioneer species adapted to colonize inhospitable, dry, disturbed ground, the seed an essential dispersal mechanism for willowherb, though more of an insurance for goldenrod, which gallops at the root. We make our brave attempts at prairie with drifts of echinacea and helenium, whose bright colours and fancy botanical names hardly suggest a connection with the coneflower and sneezeweed of their home ground; the statuesque proportions and long racemes of white flowers on fluorescent pink stems of the desirably dangerous *Phytolacca americana* assume quite the different air when it's discovered they belong to "pokeweed", and the raptures into which a Brit will descend over the seasonal interest of a multi-stemmed amelanchier will never cease to cause amusement across the pond, where the otherwise named "serviceberry trees" are a dime a dozen. Both melianthus and acanthus romp with unrestrained glee across New Zealand and the Mediterranean respectively, yet we accord them both honour and attention in our own beds and borders. Weeds, but not as we know them; liberated from that label, we seem finally able to see them for what they are.

It's this habit of ours to categorize the world around us that gets us into trouble. Ostensibly, it's a strategy to help us make sense of our surroundings – in reality, at best an excuse for lazy thinking, at worst, a tool that sows division. When we refer to "weeds", we deny a plant's individuality, conferring upon whole groups the status of "undesirable", a stamp passed down unthinkingly from one generation to the next and rarely questioned. These, though, are the rightful heirs of this land, an indigenous presence we rail against daily, trading in tools and potions for their effective repression and denial, the continued existence of a market for which is hardly testimony to their efficacy. But deep down, we know that our hold on the land is fleeting, our return to it inevitable, and the constant reminder can prove unwelcome if we set ourselves up in opposition to the forces of nature; if we view her as a thing to be subdued. In our gardens, we can work in a more collaborative way with nature than many of us have been raised to appreciate; a joint endeavour, in which we need to challenge ourselves to be more accepting of those plants that return with stubborn persistence, for the sake of the soil, the land, and the planet, yes – but also for the sake of ourselves. Because there's no weakness in accepting the inevitable, and joy to be found in engaging playfully with our surroundings, stopping every now and again to stand about and watch something temporary, beautiful, and full of life grow up around us while we wait.

To stand and stare

What is this life if, full of care, we have no time to stand and stare?
W. H. Davies, "Leisure" (1911)

Gardening has no need to involve aerobics every time. In fact, it's my heartfelt belief that the chief obstacle to enjoying your garden is all the stuff you've been led to believe you should be doing while you're there. Most of us are well trained in the idea of garden busyness, a message reinforced by articles with titles such as "Things to do in the garden this weekend". Add to this the pervasive tone of the media, the disapproval (real or imagined) of the neighbours and our in-laws, and the legacy of half-remembered gardening sessions reluctantly spent raking, weeding, or sweeping for pocket-money when we were kids. All of this fosters a notion that gardening necessarily involves work, and that this must be endured before we can be allowed to experience the joy of being outside among the plants. But what of observation? What of curiosity? What of celebrating the life already there, rather than poisoning it, ripping it out, and replacing it with something infinitely less well-adapted to the site? What if we were just to stop with all the doing for a while, stand still and have a proper look at what's going on around us? Oh, but that's hardly gardening, they'd say. But who would be delighting in their garden more, the busy bee, or the thoughtful observer? I know which one my money's on.

I am no connoisseur of poetry. I think I might have a pretty unsophisticated fondness for rhythm and rhyme that allows me to receive as much enjoyment from a well-penned (and probably slightly rude) limerick as from any rarified sonnet or romantic ode. But gardeners are earthy types, and perhaps it comes as no surprise to find us laughing loudly from the cheap seats, which I do, at every opportunity. The poem "Leisure" by W.H. Davies might not be the finest example of the form (it bears an uncanny resemblance to the mock-solemnity of "Ode to An Expiring Frog" in Charles Dickens' *The Pickwick Papers*), but it has all the necessary *tum-te-tum* I could hope for, and the message of its first two lines has stayed close to me ever since I first came across them in one of the musty-smelling, dog-eared anthologies we were obliged to work our way through at school. It was penned in 1911 and, when its creator wasn't living the life of the original super-tramp, he stayed just down the road from where I live, in the heart of the Kent countryside. The house that Davies rented from fellow poet and mentor Edward Thomas sits on Egg Pie Lane, barely a 10-minute drive from here along a fast B-road deserving of a higher classification, though I like to think a fox could make it here at a dash across the fields in barely more time. By train, it would be quicker still and, as luck would have it, the London to Hastings line runs almost door to door. Though there might be one small cul-de-sac and a few sheep between the end of my garden and the tracks, there's comfort in knowing that, were time not quite so annoyingly and unremittingly linear, Davies and I would hear the same train only moments apart. The sound of the wheels on the rails would strike a familiar percussion to the poet, their *clickety-clack* finding its way into the lilt of his verse and reminding him of an itinerant existence skipping trains across North America and Canada, and hopefully not too much of the rail-related accident that claimed half his right leg, forced the return to his native Wales, and thence to Kent via the doss-houses of London. For all the speed of locomotion, there's a stillness to viewing the land from a train, an almost meditative state that descends upon the watcher as the miles are swallowed up and the countryside reveals itself to the gaze in wide perspective, and I picture the standing figure of a raggedy Davies silhouetted against the open door of a boxcar, looking out upon the landscape as it rushes by.

It's the same kind of gaze and the same kind of rush as we can adopt in the garden, only here it's not distance that's swallowed up, but time, and the degree of movement it involves is entirely down to us and the dictates of our gardening style. Like Davies, I would advocate the cultivation of that particular state of ruminant calm the outward interpretation of which is so hard to fathom, its characteristic expression perennially labelled "vacant", though a fairer description would be "inscrutable". In such an attitude, we are very much not bothered by the opinions of others, and the state of one's body or one's features are matters of supreme indifference when attempting to become a centre of serene calm, as the rhythm of the seasons plays out all around you in the plants and the creatures and the weather, in the sky overhead and the earth beneath your feet. And the unexpected thing is, no matter how lofty or woo these aims might initially sound to the twenty-first century sophisticates we consider ourselves to be, something very much like this transformation occurs, time after time, in anyone who goes outside to fix the garden and finds instead that it fixes them. If my evidence for this is largely more anecdotal than quantifiable, it's also personal and, to that extent at least, empirical; repeatable, even, since the regular meditative practice of standing in the garden brings the same benefits to mind and body with each repetition, with no reference to the law of diminishing returns. Should you, too, experience a slowing of your heart rate, a tendency to take deeper breaths, an easing of tension in the muscles of neck and shoulders, and all of this accompanied by a degree of mental acuity that might more typically be sought through means entirely less legal, it would come to me as no surprise. Initially, such a state might seem harder to reach whilst one set of neighbours engage in loud confabulation on the subject of sweet peas and the other side applies an angle grinder to their latest project, but even these distractions become a part of the daily life of the garden; and it grows on anyway, unconcerned.

Even so, "just standing about" is antithetical to a prevailing work ethic that, for all we would deny it, still exerts its subconscious pressure; an annoying artifact of the superego that finds anything other than ostentatious activity a source of deep shame. So much of the discourse around gardening concentrates on the doing, rather than the being; but *being* itself is doing, since to exist is an act, to continue to do so with each new day one marked out by stubbornness, gratitude, and perseverance. There is no need for semaphore to justify our taking up space, despite what we may have learned, and we are privileged to have the discretion over which areas and aspects of our life might be distinguished by a degree of quietude. Supplement with an awareness of surroundings, add in observation, engagement, conversation, relationship, and there's a whole lot of *doing* going on with precious little to be seen from the outside. Placed in the context of a garden, we may well find that we spend a great deal more time than we'd initially imagined apparently just standing about, gazing off into the middle distance. But in moments like these, and quite contrary to appearances, gardening is still going on.

"A garden needs to be cultivated!" The old gent is unimpressed by the garden that's just taken the coveted "Best in Show" award at the Chelsea Flower Show, and he's happy to let one of the judges know what he thinks. The trend this year leans strongly towards a naturalistic style of planting – one which, looking up from where I work, I can see all about me in the glorious, rain-splashed tangle of my own garden on an overcast summer's day. It's not a look to everyone's taste, and it's clear that Lulu Urquhart and Adam Hunt's masterly depiction of rewilding beaver activity and natural habitat restoration in their "A Rewilding Britain Landscape" garden isn't cutting the mustard with the judge's new friend. Here, clearly, is a gardener sure of his ground, confirmed in opinions which he has every right to hold, and with which others have equal right to disagree. He expects – perhaps not unreasonably, given his evident vintage (to which we must accord respect) – the hand of the gardener to be more apparent, even in the absence of the gardener themself.

He wants, for the space to qualify as a garden, to see indisputable evidence of cultivation, from which I think it's safe to infer, he means crisp edges, neat lawns, and orderly ornamentals, rather than the unruly sprawl of cow parsley, bistort, and foxglove bordering the stream's edge in this powerful little vignette.

But what exactly does it mean, to cultivate? We cultivate friendships, understanding, and identities – almost half of us, at one point and with varying degrees of success, will have attempted to cultivate something upon our top lip – but none of this activity requires spade or lawnmower. Quite simply, we place ourselves at pains to grow something that wasn't there before, and whether it looks drastically different or remarkably similar, our old friends intention and purpose have made the change. In the garden, this plays out first with the declaration: I *intend* for this space to be so; and then, with the investing of the space, the plants, our own efforts, with the *purpose* to achieve that end. Two aspects of will, the one cerebral, the other perhaps more active, though not necessarily physical, and eminently capable of being occluded from outside observation by the busyness of a beaver, or any other of nature's creatures.

For all that it's a construct, a garden straddles both conceptual and physical realms and, being in possession of that innate ambition to grow itself into a state of wilderness, it would be fair to ask if there's anything of practical sense and strategy behind all this perpendicular contemplation, or whether such a laid-back approach can only end in the gardener gradually disappearing beneath a turbulence of bindweed and ivy. There is method to the madness, and a sound justification for staying the gardener's hand that has at its root that general principle explored in *A cut above* (see page 170), where we discovered that a plant's response to being cut is often to grow away with renewed vigour. Extrapolated across the whole garden, and borne out by observation, it's not unreasonable for the gardener to form an opinion encapsulated by the excellent tenet: *The more you do, the more you have to do*. The more we mow the lawn, the more we need to keep mowing the lawn to maintain the close-cropped manicure of the grassy sward. And not merely to mow, but to water, weed, feed, rake, roll, and scarify, if we truly want it looking its green baize best. There may even be stripes, and they don't get there by themselves; no, nor linger, neither. The more we weed a flowerbed, clearing a gap around our prized plants to give them the chance to flourish without competition from or reference to the true children of the soil in which they're plunged, the more we will be required to keep weeding it, if only to maintain that appearance of sterile no-man's-land between one specimen and the next, exposing germinating seeds to the light with every cultivation. The harder we prune a deciduous shrub or tree at certain times of year, the harder it will grow back. The more we cut back plants and cart off their remains to the compost heap, the greater the need for mulches to maintain nutrient levels and soil texture. The more we do – the more we bring into the garden – the more resources we burn in time, energy, and money; all those trips to the garden centre in the car for potting mix, chemicals, and widgets. All those litres of waste sent out for the municipality to process when we could be returning their goodness straight to the soil. Thanks to a significant portion of the gardening lore passed down to us from one generation to another and perpetuated by the structures which surround us, we are literally making work for ourselves, while helping the planet to fry.

The corollary of this is that the more we let nature have her way, the less time we have to spend making amends for the mess we've made of things in the name of tidiness and horticultural excellence. More tolerance of the natural flora of our landscape means less time spent weeding, while longer, shaggier lawns with more wildflowers for pollinators support greater biodiversity and fewer trips out of the shed for the mower. More latitude extended to plants to stand about looking less than perfect, the richer the ecosystems our gardens can support. More time spent standing and staring – apparently doing next to nothing – means more notice being taken; less head-down hustle means more opportunity to observe where improvements could be made, or to catch a poorly plant before a minor issue has the chance to become more terminal.

I'll be the first to admit that there's more than an element here of post hoc rationalization for an attitude I find myself naturally inclined to strike. What can I say? If a long-held inclination for lucid daydreaming has opened my eyes to the benefits of a pause and a ponder, it would be selfish not to share the discovery. But for a pause to be, there needs to be activity on either side, and I have found that beyond its immediate function, this busyness has a value on which I have come to depend. It is a universal phenomenon, and it is this: that however small the space in which you garden, it will be big enough to swallow up both you and all your worldly cares; a doorway to the rhythms of the natural world in which you'll find the details of your life hold scant significance. There is comfort in such humbling anonymity, in recognizing ourselves as a tiny but intrinsic part of something whose vastness we can't begin to comprehend. I have fed the garden with my time and energy, but also with my feelings; watered it with seaweed extract and compost tea, with weariness, grief, despair, confusion and, perhaps as a result of time spent out here, not infrequently with thankfulness and joy. This soil and the life it supports has sucked up my sweat and my tears (of sadness and laughter) in equal measure, and thanked me with, in addition to flowers, a refreshing indifference. This has come as one of the most unexpected but significant benefits of what I understand as "gardening" and provides an essential counterpoint to the stillness I seek to conjure whenever I set foot on the garden path, a seesaw of truth that lends meaning as well as mud to the endeavour. We can lose ourselves in busyness. But in moments of thoughtful engagement, we find ourselves again.

True to his itinerant nature, Davies didn't linger over-long in Kent, eventually moving back to London, though the list of his subsequent addresses suggests that he never truly settled, and he died in Gloucestershire in 1940 at the age of 69. I find myself wondering if only those who rove most widely can truly appreciate the benefits of taking the time to stand and stare, yet I am no wanderer by nature and, rooted to this ground, had incorporated the pose unthinkingly into my daily practice for who knows how many years before I realized quite what I was doing, or whose example following. That space we carve out of our days for thought and contemplation is experienced most keenly not so much in response to the miles we've travelled, nor the distance covered bustling in frenzied spirals around a small plot where our beginning and our end are just one pace apart, but as a contrast to a busyness of mind that thrusts us ever headlong into the rush. It can be good to lose ourselves in the melee; to expend our fraught and frantic energies in the furtherance of some goal, to wrestle with the messy middle and emerge victorious on the other side with achievements we can count in place of sheep when the time comes for laying head upon pillow at the end of the day. The garden offers generous potential for quick wins and short-term accomplishments. But the best gardens, like the best friendships, need to have lavished upon them thoughtfulness as well as effort, and the moments spent in quiet companionship with our friends are never lost.

Summer draws to a close as the busyness of the garden subsides, growth and pollination largely achieved, and every flower viewed now as an increasingly precious gift. We deadhead in denial of the passing of the year, dipping down towards autumn with golden afternoons, blue skies raked with the feathery plumes of grasses gone to seed and displaced daisies from distant lands too confused to know better than to bloom, though they're not alone, the flowerbeds full of bumblebees clambering over sedums and salvias or reversing out of the long tubular corolla of a beardtongue. To gratify our magpie eye, we extend the flowering season with colourful introductions, though kneeling into my work with head bent low, half in and half out of a border, it's easy to get lost in the detail of doing and overlook the full implications of something that's become second nature. It takes a little distance and a wider perspective to see that there are two gardens on the one spot, though both moved by the same promptings – to grow, to live, to multiply – and surely the life of the gardener can only be made more easeful by the recognition that the key to their role lies in working out how best these two gardens, the one belonging to the land, the other super-imposed upon it by those humans presently in possession, can be united around this common purpose. It's something worth a few moments' cogitation. A time to stand and stare.

How to STAND STILL

The wedge of pages in your right hand is growing ever thinner, you're eyeing up the next volume in your To Be Read pile, and still I'm yammering on about standing around in the garden. It's kind of central to this whole thing. And part of that is because what I'm getting at is so easily misunderstood.

This posture is not so much a rigid and physical affair. To stand still here, with purpose, is not about the absence of fidgeting. It's not even about standing – some may need to sit or lie instead, and this should present no obstacle.

This is about entering a state of receptive restfulness, immersed in your surroundings and alive to every detail, while feeling no immediate obligation to make a change. It's about being engaged, and taking notice, and maybe thinking about sorting this or tweaking that… but also, maybe not.

So, fidget, flick your fingers, wander, or wheel yourself about a bit. If that's what being at rest looks like for you. Whatever you need, to feel truly present.

Go into the garden and find that spot where you feel at once most comfortable and, also, most alive. You will know the one.

A pause. A breath.

Do nothing – your own best version of doing nothing. This is harder than it sounds.

Now, slowly, begin to cycle through your senses; scent first, then taste – the air, a blade of grass, a hawthorn leaf – then touch, and hearing. Leave sight till last, we tend to overuse that one if blessed with it. Become adept at concentrating upon signals from each source, and then in combination.

Take a mental inventory of what the garden contains: plants, animals, and birds, sounds and smells. This is never complete but becomes more comprehensive every time.

Cultivate a habit of passing the time of day with anything on the above list.

Offer up thanks, in whatever manner and direction you feel led, for the fellowship of soil and sky, leaf and flower, bird and bee, before carrying on with your day. And resolve to come back and do the same tomorrow.

⁕

Look, I'm standing, you say. Standing still. Can I get on with things now? Well, yes of course – always! But also, maybe hold up a while. What's the hurry?

FRUITS

to the gardener

We have arrived. And all the industry of summer slows
to store its golden rosy light in myriad expectancies.

And some will swell, and some will split, and others
dry and wrinkle on the stem; and some will fall, and
some will fly, and some will make their journeys in the
company of fox or thrush to contribute their excellence
to communities undreamt of on this soil. The tired
pollinator comes to rest. The busy daughter spins and
strikes and sucks, recycling redundancy; for nothing –
as you know – will ever go to waste.

But you and I have games to play. The earnest lines you
draw, the heaps you build – precarious before the fickle
autumn wind. It's too much to resist. In recompense,
I'll send you home with apples, blackberries, and sloes,
and pockets stuffed with nuts. And we will do this all
again, and not just this, but all that other, too; and much
of it will seem the same, and much else, somehow, new.

Ripeness to the core

Season of mists and mellow fruitfulness,
Close bosom-friend of the maturing sun;
Conspiring with him how to load and bless
With fruit the vines that round the thatch-eves run;
To bend with apples the moss'd cottage-trees,
And fill all fruit with ripeness to the core;
To swell the gourd, and plump the hazel shells
With a sweet kernel; to set budding more,
And still more, later flowers for the bees,
Until they think warm days will never cease,
For summer has o'er-brimm'd their clammy cells.
John Keats, "To Autumn" (1819)

Lunchtime, and the sun is still trying to burn off the morning haze. Slight breeze, russet tones, mushrooms. Smell of wood smoke and the burnt-sugar tang of leaves on bonfires, like an unattended candyfloss stall. Autumn. Distilled down, these are among the ingredients that would yield the essence of the season; a time of year which would be perfect were the days that bit longer. But it's this very same reduction in daylight hours that plays a crucial part in creating one of the quintessential features of autumn, as the production of chlorophyll slows down and, in the leaves of deciduous plants, greens fade to rich oranges, yellows, and browns. It's odd to think that these carotenoid pigments are present all year round, but that only now, without the masking effect of the chlorophyll, do we get to enjoy them in all their glory. At first, this change takes place in the vertical plane, as trees and shrubs extract those nutrients from leaves to store over winter in the permanent framework of stem, branches, and roots. Then, no longer of use and containing only an unwanted trace of sugar and other carbohydrates, the leaves begin to fall, gradually occupying a more horizontal orientation, until the whole world seems carpeted in opulence. Reason enough to venture outside – welly-clad, shuffling through piles of crisp, spent foliage. Or armed with a rake, creating lines and heaps for gathering up and cramming into bags which in 12 months' time will contain the best, most crumbly soil conditioner imaginable. But that is some way off. For now, we have leaves to tame. And to steep ourselves in autumn before winter robs the world of its colour.

It's the light that marks the shift. Perhaps too the morning mist hanging in heavy layers over the fields and creeping through the streets to settle in our gardens, but these come quietly in during August and belong as much to summer's end as to autumn's beginning, a veil drawn across the threshold between one season and the next, effectively obscuring the point of transformation. It's a messy join; one season steals in upon the other – now autumn in the hedgerow, still summer in the field, or vice versa down the way – and every plant reserves the right to mark the change according to its own schedule. But one day, the light alters as though a switch is flicked, and we are on autumn's time; the stark, overhead glare of summer sun replaced by more mischievous and creative rays that bring a warm, gold, radiating glow with bright and argent highlights. Where summer baked and bleached, autumn intensifies, colours saturate, and flavours layer their richness one upon another.

The mood changes with a plant's priorities, all need for seduction abandoned with the last of the petals, both breeze and bees have served their purpose and, heavy with expectation, the garden shifts in focus from flowers to fruit; even more so when we let it. Our deadheading activities, indulged in to garner blooms for the house, to maintain the tidiness of a shrub rose or deliberately to prolong the flowering season of a plant, push the appearance of certain fruits back in the year. Hips appear much more quickly on the dog roses of the hedgerow than on the cultivated specimens in our borders that have received our rigorous attentions, and the floral display of our annual sweet peas would be over almost before it had begun were it not for the conscientious snipping away of every nascent pod the moment it appears. All, in time, succumb, and even the colourfulness of the asters is transformed as the neon-bright mountains in the late border are replaced with clouds of cotton wool, each bright-eyed daisy a white fluffy button consisting of a launchpad crowded with fruit, every floret transfigured into a precision-engineered seed dispersal system. The fruit of a daisy – any composite flower from a dandelion to a sunflower – is not a luscious thing, but a dry, non-splitting (indehiscent) case that encloses the seed and known by the eminently forgettable name cypsela. Topped, as in the case of the aster, with a hairy parachute or pappus (meaning "old man"), which might be clutched tightly to the fruit, or held upon a stalk, the means of making its way in the world are conspicuously evident, the plant having relied upon insects for pollination nevertheless resorting to the wind as a convenient measure for the distribution of its progeny.

The theme of distribution and dispersal holds the key to the season, though it's not the aspect of autumn that springs most readily to mind. But the development of the fruit around the seed – the ripening of which informs our own understanding of harvest time due to its importance for our own food supply – is really little more than a means to an end. If, the better to understand how this gardening thing works, we're to continue with the notion of thinking like a plant, then some effort ought to be made to understand the importance of fruit when viewed from within the flowerbed (see *Going to seed*, page 248). Keats was at least halfway there when, for posterity, he yoked the season of autumn so closely together with the behaviour of plants and their pollinators, but he only allowed himself to get partially botanical in his descriptions before being borne away upon the wings of poesy. The rhapsodies of "To Autumn" stop short of examining the function of fruit, presenting us instead with a charming version of a familiar and highly anthropocentric narrative; the fruit, as the honey in the bee's overflowing and clammy cells, is for us. Yummy.

For millennia, autumn has been synonymous with harvest, a relationship that increasing urbanization, the global reach of the supermarket's procurement processes, and the obfuscation of seasonality in our weekly grocery list have yet to render entirely obsolete. For all the efficiency of the systems which stock the shelves of our grocery stores, there remains within our collective consciousness an impulse to give thanks for the blessing and abundance of the fat years, and to pray we might be spared from the threat of want and hardship in the lean. As individuals, the more in touch we are with the food on our plate and the land from which it springs, the more readily gratitude rises from within us in response. It's right that, especially at this moment in the year, we celebrate the fruits, berries and pies, the crumbles and cobblers that will follow, the leaves and roots that we'll win from the vegetable plot, the corn and wheat from the fields. This is the food that will feed us through the tough, dark winter months and into the new year, while the soft, still warm soil into which we sow seeds, and the gentle rains that water the ground will give our spring crops the best possible of starts. But all of this – every plant that we harvest and every seed that we sow – is only possible because of this timely state in their lifecycle, a culmination of months of effort expended in rooting, growing, flowering and fruiting and – though we might manipulate each of these processes to our own ends – all intended not for us, but for the furtherance of their own agenda, the continuation of their genetic legacy. We can talk of reciprocity and kinship between humans and plants, and I truly believe these exist, but at the same time it's hard to escape a feeling that we're simply included in these enterprises under gracious sufferance, and that our absence wouldn't be missed for more than a season or so.

And so autumn, viewed from a plant's perspective, has nothing to do with what gets placed in my fridge or stored in your pantry. Neither does the garden view the season as a cool, crisp (more often damp and soggy) addendum to those long, hot, dazzling days of summer full of garden memories and dreams. Instead, autumn is the high point of the year to which everything has been building all along – as far as the rest of nature is concerned, this is peak plant, and the garden arrives at its destination just when we begin to lose interest in it for the year.

I mean… I say "we" – it's the prevailing behaviour I recognize in those around me. To an extent, we're all a little guilty of it – that habit of returning from summer holidays and maybe squeezing a few more sunny weekends out of the garden for barbecues and long evening get-togethers before it's back to work and back to school, and even the fire pit proves insufficient to banish the growing dampness and chill on the evening air. At this point, the garden becomes relegated to the status of spare room, albeit not a particularly useful one for the next few months, having no roof and a propensity to accumulate its own clutter. Still, we can shut the door on it, all the same. Fair-weather gardeners, you see; I called it back in *Wonderland* (see page 18).

There's a strong argument that good garden design can solve many of these issues. Hard-wearing, nonslip surfaces, weatherproof furniture, a clever layout that acknowledges the issues inherent in connecting indoor and outside space together whilst creating a sense of synergy between the two. The difference a talented designer can effect upon the way in which we use our domestic space is transformative, but the commissioning of such magicians shouldn't be the sole prerequisite of enjoying autumn in the garden. Sometimes, all that's called for is a thicker jumper, a hat, and a refusal to be cowed by inclemency. There's too much to miss to let a little weather get in the way.

This is the time of burnished brights and smoldering, dusky blushes; of low sun and silhouettes; of backlit borders and fleeting otherworldly moments as the sun rises and sets, bathing the scene in warm, golden light that fades while you stand watching its progress across the garden, and long before you've had the chance to tug your gloves off with your teeth and retrieve the phone from your pocket, or the camera from its bag. And, on the days when the sky remains disappointingly blanketed with cloud, there's a luminosity to moss, grass, and ivy and a chocolate fudge richness to the brown of bark and soil you feel you could sink your teeth into, while receiving nothing but healthful improvement from the experience. Onto these surfaces, the first of the falling leaves scatter themselves with an artistry and finesse suggestive of the painstaking attentions of some unseen hand, cast bronze and beaten copper accents upon the soft and yielding, greenly glowing ground. These are the outriders of the great multitudes yet to descend and, finding the ground largely unclaimed, they capitalize upon their free time by cavorting with every gust of the autumn breeze, before being joined by rank upon tumbling rank of their comrades to lie in drifts and heaps, awaiting further orders. It is necessary for the gardener to make their peace with leaves, from which there is no autumn escape, demanding to be dealt with in all but the wildest of schemes.

If any kind of regular presence is to be maintained throughout the season, I've found it by walking face first every morning into at least two spider webs slung overnight across the garden path, and a further one on entering the shed or the greenhouse. I feel hugely guilty each time I destroy such intricate work, though it's nothing to the sinking feeling that comes on hearing the crunch of a snail underfoot. For all my internal resolution to dance, I fear there's a lumbering quality to my movement about the garden, and incidents such as these provide a salient reminder that, while my participation out here is tolerated – perhaps even, I like to think, welcomed – it's not without its consequence, intended or otherwise. Rather than beneath the heavy sole of my boots, or as a result of ingesting pellets or even nematodes, I'd prefer any surplus mollusc life in the garden to meet its end as dinner for a hungry thrush, frog, or hedgehog, and while a bird can easily avoid my less than elegant flailings, the responsibility to look out for ground-dwelling creatures who might now be turning their attentions towards hibernation is keenly felt. The damage that can be wreaked when poking about with a garden fork is one thing but, along with carelessly-managed bonfires, power tools pose the biggest threat to garden wildlife, and none more so than the nylon-line strimmer – a tool that excels in dealing with just the kind of post-summer overgrown tangle in which small animals seek shelter. Here is one surveillance operation where, contrary to common practice, the best approach is to make as much noise and disruption as possible, to scare away creatures hiding within. It doesn't take long to scope out an area before going to work; a tap and a prod at the ground to be cleared, a moment's careful attention for the sound of a light scampering, or the glint of a fearful bright eye. For the same reason, a fire is never lit where the materials intended to feed it have been stacked up in storage, but a new pile assembled each time, upon clear ground and with sufficient and deliberately oafish disturbance to rouse even the sleepiest of creatures and give them leave to seek less imperilled shelter. We should be in no doubt, despite our kinship, that the presence of even the most benign of our kind upon the land does some kind of violence to the life with which we share it. In the garden as elsewhere, we must do what we can to soften each step and every impact, to make each cut with surgical precision and in such a manner as to stimulate further growth or, at the very least, in some way to nurture and preserve life. And this is not to suggest any dichotomy between humans and the natural world; for this chapter of our planet's history, the one is part of the other, with no allowance to be made either for anthropomorphism or pathetic fallacy, both of which presuppose a separation whose existence, though often claimed, remains unproved. Rather, it's an unequivocal acknowledgement both of our place in nature, and of the impact we have upon those in whose company we exist. As so often, we wound those we love, but we grow together anyway, and our scars tell the story of the time spent in each other's presence.

Autumn is traditionally a time of slash and burn, from the stubble in the fields to the ever-growing piles of spent foliage and stem in the garden and, while much of this is bound for the compost bin or shredder, the fires we light serve not merely to help us process the leavings of the year, but to push back the growing dark. If, in these days of poor air quality and respiratory disease, that fire exists more in memory than in practice, it retains its currency as a beacon of change, a milestone in the year marking shorter days, colder nights and the slowing of growth rates in the garden that such a particular combination of factors inevitably portends. A change of pace, but also of focus, as the gardener raises their gaze from what can be achieved this year to what might be accomplished in the next. There are decisions to be made in the here and now – what to cut back, what to cut down, and what to leave standing – and warmth enough still in the soil to accommodate the growing roots of new plants, or of the divisions of old. And while undeniably for many of the garden's energies autumn represents a time of drawing down and in, it also calls time for the collection of seeds for the new year, and the sowing of the first crop of the spring to come.

Going to seed

This is what it's all been about, all along. You thought the garden was for you? Fool; think again. You've been duped, but what a ride. She agreed to have you along, used your brawn and, maybe, a very little of your brain, but all the while the aim was less to create for you the perfect garden, but secure for her the next generation. Do you care? Not even slightly. Same time next year? See you there.

Sometimes it can feel a little like this when autumn rolls around. That the garden I want to create and the garden that nature is intent on being in this spot might occupy the same space, but when it comes to sharing the same purpose, there's still a way to go. I don't ever set out at the beginning of the year with the explicit intention of having a garden full of seeds, though, looking around me, this is exactly where we've ended up. And if I think a while about what exactly that means, I begin to wonder why that hasn't been listed among my priorities for this space all along, and start to entertain ways in which it might be in the future.

Perhaps part of the reason behind both the endurance of the human-nature partnership in the garden – also some of its tensions – is that we share with plants a similar concern regarding what's left behind after we're gone. Our existences coincide, but we are moving at different speeds and to different timescales. And so, while the gardener could be excused for not wishing to dwell overlong on what comes next, for plants the prospect is a question with an altogether more pressing sense of urgency. Still, there might be 70 lives of an annual plant to one of ours, which can seem like an enormous disparity but, from a geological timeframe we're in this together, our importance and our insignificance bound up within the soil to which we'll all eventually return. Why wouldn't we stand shoulder to shoulder with one another during a journey so fleeting that we barely have time to snatch meaning out of the flow of the passing seasons, but where the reason that does lie within our grasp finds its fullest expression in engaging with life in all its fullness, in pouring ourselves out into every stage of existence and in nourishing our fellow travellers along the way with the product of our efforts? To spend our days in the company of plants and the myriad lives which they support (including our own) in their inexorable quest to grow and multiply and keep a toehold upon this earth – and if that involves growing old together, and watching over as the other goes to seed, so much the better. We will count it a privilege or be damned.

Looking back upon the year, we realize that plants have been packing copies of themselves into tiny capsules ever since the first flowers of spring. Some, the little hairy bittercress and other ruderal things whose survival is predicated upon the strategy of bombarding with a fusillade of seed the frontier ecosystems in which they slug out their tough existence, will produce several generations over the course of a single season. But autumn is where it all comes together; the time when, with very few exceptions, the journey from flowering through pollination, fertilization to fruiting and seed production ceases for a few months over winter. But there is something in our language that would rob this culmination of its moment, a disparaging tone that links seediness to weediness and so successfully alters the meaning of both that we rarely pause to question what we're saying when we employ them in a phrase. To introduce a note of seediness is to entertain the disreputable, to exhibit any aspect of weediness is to expose a lack of strength and resilience, while to "go to seed" offers greater cause for lament and pity than celebration.

But weeds, despite their raggedy reputation, are ideally suited to their environment and when left to their own devices, are adept in fulfilling a destiny that incorporates flowering, pollination, and the production of the next generation. Rather than a contra-indication of the same, it's a facet of their success that, during so many of the occasions in which we can be bothered to notice them, they are arrayed in a state of seediness; and if they have had to take a blow to the health and vitality of their own bodies in favour of their offspring, what caring and responsible parent has not done the same? Plants produce seed and, in doing so, have the life sucked out of them – entirely, in the case of annuals, to a lesser but still noticeable degree with perennials. There's something entirely unsavoury in the knowledge that these aspects of a plant's lifecycle inextricably linked to parenthood – specifically, to *mother*hood – have become enshrined in our vocabulary as bywords for weakness and ill-repute. Language shapes the way we think (it's why invading powers have repeatedly sought to establish control by eradicating the indigenous tongue), but the way we've been thinking undermines our supposed commitments to individual mental health, the cohesiveness of our society, or the wellbeing of the planet. If we're to have a hope, we need to reclaim the notion of being seedy and weedy from those who seem terrified of the reproductive power of nature. We could start by wearing it as a slogan across our T-shirts and chanting it in our streets, as well as embracing it in our gardens.

*

And so, the garden goes to seed. But stand with me a while and take a reading of this time and space. We have the measure of this now. *Breathe in, then hold; now, slowly, out*, one essence mingling with countless more, and we are in the garden and the garden is in us, and energies align. This whole… *whatever it is…* is becoming less awkward. Less difficult to discern too, the mood of the garden; the better we get at slowing down and listening to ourselves, the better we become at sensing shifts of tone and emphasis in others. It's a harmony, a vibe, and while it's taken practice to understand how best to downshift from the day into a more attentive state, it's hard now to remember our earlier disconnect. We look around at drying stems, yellowing and corrugated leaves, *warp of dried raffia, weft of crinkled biscuit* and see the garden start to slump into itself. Heads begin to bow and shoulders droop – it's fine, it's good, it's all to be expected. If there is tiredness here, it is not the sombre weariness that comes before a death, but the burn of fatigue after strenuous exertion; the exhaustion that comes on completing a marathon when every muscle aches and limbs too heavy to lift seem curiously reluctant to obey instructions; when you know that you will do it all again in time, but right now you must rest. A happy state of being temporarily spent, in service of a cause. A rhythmic lull, *une petite morte*, where satisfaction is followed in short measure by a sleep. So, gone to seed, but no question here of being past it. Merely, finished, just for now.

*

For all a gardener's delight in the discovery of ways in which to influence the behaviour of a plant, there are limits to our ability to put a pause upon the year. Deadheading faded flowers has drawn out the display, a mindful occupation all summer long, but with an air of inevitability the rose turns its thoughts to hips, sweet peas to pods (they've been threatening this for months), and sunny cosmos blooms and fades before we have the chance so much as to brandish our secateurs. To everything there is a season; the year has turned, and flowers are reworking themselves into something rich and strange as the fertilized egg transforms itself into a seed, and the tissue of the surrounding ovary envelops it with fruit of one kind or another: a swollen, fleshy berry; a hard-walled nut; a dry, single-chambered vessel with a parachute attached to name but three. And whether the wind is gladdened by the lightweight helicopter wings of the maple, elm, and ash that launch their seeds onto its breezy autumnal gusts, who can tell, but there's no denying their ingenuity of engineering or the specificity of form, and no room for doubt when it comes to the function of fruit.

For some plants, having dropped the floral finery months back once the seduction of pollinators had been achieved, there's now once more a job to do in order to attract the right sort of help, but the way in which nature suits intention to purpose is never disclosed more candidly than at this point in the year, when the bespoke tailoring of the seed's first container to its distributor becomes apparent; the acorn to the squirrel, the berry to the bird, the samara to the breeze. There's a message to be gotten out and broadcast far and wide, a genetic code to strengthen present populations and form the basis of new communities. Plants rely on a combination of specificity, opportunism, and reciprocity to disseminate the word, and they shape their fruit accordingly: whether to fly like the sycamore on the wings of the wind or to shoot their contents out with the unexpectedly explosive force of the violet; like the coconut palm, to drop it into the ebb and flow of the ocean's tide, or onto the current of a river, like the alder (or the Himalayan balsam, which combines two methods of dispersal, and shoots its seeds into the water courses on whose banks it habitually grows); to hitch a ride on the outside of a passing creature by clinging to its feathers or fur with the barbed burrs of avens or burdock, or to travel through its gut and out again, smuggling the seed into its living transporter through the kind of "Trojan pudding" method employed by all plants that draw animals into their service by the tantalizing prospect of energy-rich berries, and delicious fruits.

It's onto these cases of mutualism that we can be accused of projecting our own notions of the bounty and provision of the natural world and, though it's hard to ascribe any motive to nature, we'd be as foolish to deny it as we would to claim its acquaintance. We decide, then, what to believe, and if today I choose to play down the transactional nature of these encounters in favour of a narrative that focuses on the generosity of plants in sharing their abundance with those creatures, including us, with whom they share a space, that's all well and good. Cold objectivity and calculation have their place, particularly when there are decisions to be made, but we understand our world through stories. I like the ones about nurture and benevolence, and most often when I walk out of the back door and into the garden, it's into those stories I'm looking to step. Nature, Janus-faced, embodies either aspect with equal conviction, but for me, there's always tomorrow for realism. Or the day after that.

The currency of the autumn garden undergoes a conversion, the value for creatures held in sweet, pithy flesh and nourishing endosperm, for the plant in its embryonic offspring. The coin, in either case, is the same, transforming from flashy floral exclamations into more modest tokens, though jewel-bright shades to catch a passing eye are not uncommon among the dun and the brown. It is a time of fruit and berry, nut and seed, and we know what we mean by the terms, though botanists might demur. It can be useful to know which part of the flower became what part of the fruit and have terms like "drupe" (stone fruit; the plum family) or "pome" (apples, pears, and quince) to throw about but, call them what you will, they'll all taste just as good beneath a layer of crumble and swimming in a yellow lake of custard. I can be temporarily entertained by the account of the aggregate of drupelets, each with its single seed, that comprise a single blackberry, but I'm much more interested in cramming them, running with rich crimson juice and still warm from the sun, into my mouth, and just how to get to the juiciest fruits that are always slightly out of reach. If it's possible to extrapolate from my own obsession with fruit to a general opinion about a plant's strategy for distributing its seed, you'd be hard pressed to consider it anything other than highly effective.

But there is wonder enough to the subject of seeds without having to smuggle them into consideration smothered in sweet and unctuous juice. There is magic in miniaturization, in the ability to cradle the code for each plant in a border in the palm of one hand, to carry the secrets to an entire gardenful in a pocket. More prosaically there is convenience too, for the garden can be dried out and stored away for posterity, or transported, its several parts gifted, traded, or brought upon journeys freely taken or made under duress, every seed cramming within its case a memory alongside its slumbering, embryonic charge. Or, arriving at its destination unencumbered by backstory, providing the potential for new tales to be told: "*Alchemilla mollis*, great clouds of lime green flowers that brighten up the garden path; blue leaves that preserve raindrops for inspection. Give it a sunny inch, it always takes the mile. Beloved by bees and rain and toddling Jen, who always cried in August when it turned brown." Seed envelopes – the brown paper kind you buy in which to store your own saved seed – need to be large enough comfortably to record this kind of detail. We don't all get a hard shell into which to cram the important stuff.

There is fascination too in the variety of form, inherited no doubt from the architectural display of their containers and dispensaries: witches' hats left behind on echinacea, plucked free of seeds by birds; round and fluffy buttons of ligularia, pale ecru-coloured pappi in deep contrast to liquorice-chocolate stems; skyrocket chambers of Turkish sage surrounding felted, fluty central shafts; the poppy's famous pepper pot; the weird and wondrous curlicues atop nigella's striped balloons. And seeds as tiny as those of snapdragons, or chunky as broad beans; the knurled clockwork wheels of hollyhock, each a tiny gear within some intricate device; the spiny sea creature crescents of calendula, the angular vertebrae of beetroot and Swiss chard; every one demanding close and careful observation, and applause sufficiently gentle so as not to disturb them in their rest or send them tumbling in a jumble to the ground. And Lord forbid a sneeze unless we're keen to leave our planting plans to chance, to broadcast next year's garden in random order and design.

For such is what the season's all about, or at least what plants are trying to tell us if only we would hear their voice above the noise of us writing off the garden for another year. For all the fuss to get sowing seeds in spring, a good portion of the garden can be started now, with winter hardy annuals employing the days before hibernation settles in to develop beneath token overground presence subterranean structures of greater magnitude; a root system that will allow them to steal a march on their spring-sown companions. Frothy, white umbellifers of ammi and orlaya, golden Californian poppies and English marigolds, sweet peas in every shade, and cornflowers in royal blue and purple-black, all flower sooner and longer when sown just as the garden goes to seed; and those things that, sown now, won't germinate immediately but stratify in winter cold and spring to life with longer, warmer days – statuesque angelica, foxgloves and achillea, magenta and white centranthus and the spindle of the hedgerows – sown now into a pot and left inside a cold frame or unheated greenhouse to keep fridge door shelves free for condiments and cheese, will likewise contribute their splendour to the flowerbeds while making far fewer demands upon the gardener than anything sown in the new year. More than this – and here's a thought – we can let the garden sow where it will, and trust to the prodigious self-seeders for the display.

Ironically, it may take an effort of will to surrender our illusion of control and allow plants to let their seeds drop naturally to the ground – *one for the rook, one for the crow, one to wither, and one to grow* – and we will need to check any impulse that might lead us to be over-zealous in our clearing away and weeding. Tidiness and spontaneity make poor flowerbed fellows, and many a dreamt-of colony of hellebores, nigella, or *Verbena bonariensis* has been inadvertently forestalled by too vigorous a clear up. Here, again, we would be advised to let it go, and let it grow. And nowhere, at this time of year, is this more important than in the meadow.

Into the long grass

We're all looking for the win-win these days – those happy resolutions where mutual benefit accrues to all involved. Giving your lawn over to a wildflower meadow, in whole or in part, falls squarely into the drop zone while managing to conjure images of pastoral idylls in simpler times – hazy summer days, wandering through long grasses and tall flowers, surrounded by butterflies and the buzzing of bees. The lazy lawn, perfectly suited to all for whom the notion of pacing up and down behind a mower belongs firmly to a previous generation (along with Brylcreem and tank tops); the double victory arising courtesy of a low-maintenance management regime that actively encourages increased biodiversity.

But low-maintenance is not quite the same as no-maintenance; autumn is here, and for one final time this year, the meadow needs mowing. It's time for the hay cut.

This is not the way most of us have come to understand grass. For those with neither an agricultural nor deeply rural background, grass is synonymous either with lawn or sports turf, the close-cropped field of play that forms a venue for childhood games in the garden or contests of even more excitement on the football pitch, on Centre Court at Wimbledon or the fairway at Wentworth. It's the green baize carpet that quite literally underpins summer picnics in the park or perhaps, leaving the city behind us, the privilege of lazy afternoons watching cricket matches on the village green, while we pick absent-mindedly at the daisies to a gentle soundtrack of leather on willow and the occasional hopeful "howzat!" The meadow, as a concept, is loosely grasped – much more so than the lawn mower, whose form and purpose are clearly recognized and understood from an early age. Grass is supposed to be green and short and, when it isn't, there's something wrong that requires putting right. We are trained to be distrustful of grass when it gets long, counselled against letting it grow under our feet and encouraged to view it as a metaphorical wilderness fit only for hiding away the worst of our notions and projects, which we kick into it in the hope that obscurity will breed forgetfulness.

We might look out upon areas of grassland as we speed by them at the side of transport routes, on the train or in the car and, particularly in the case of the latter, we appreciate instinctively one of the defining features of grassland – that, in order to exist, it requires management. In the public space, conflicting values come into play around the subject of cutting grass: there will be aesthetic considerations, which are a matter of taste; environmental considerations, which are a matter of science; and practical considerations, which are down to common sense. Whether you or I are enchanted or dismayed by the look of tall grasses and the wildflowers that make their home among them, there can be no argument that reducing the frequency of mowing supports greater biodiversity, from ground-dwelling invertebrates, to flying insects, to the birds that feed upon them and the seeds that will be produced as the plants ripen and mature. But if the long grass of a roadside verge on an awkward corner obscures the view of oncoming traffic, it's going to need to be kept lower to avoid accidents. Like all plants, grass has a habit of growing, and we have to come to some decision about what we're going to do about that.

Within the biome of the temperate deciduous forest, grassland is a condition from which landscape is eager to sprint away, the vegetation transforming quickly into scrub through the incorporation of taller, more enduringly woody plants, before transitioning ultimately into woodland. Historically, areas of grassland have been held in their characteristic state through being managed either as pasture for grazing livestock, whose presence on the land guarantees close cropped grass while denying shrubbier growth the opportunity to establish, or as the kind of meadow that gives grass the licence to grow tall before being harvested for hay upon which to feed the oxen and horses that pull the plough. Since it's almost impossible to separate agricultural practice from the existence and preservation of this type of vegetation, it comes as little surprise that changes in farming, combined with pressure upon land use from home building, industry, and transport, have led to a massive decline in grassland areas, with the UK alone having lost 97 per cent of its meadows within the last hundred years[19]. Ancient grassland, with its rich diversity of plant and insect life, is encountered today in a fraction of the pastoral locations where it once flourished, as well as in a scattering of well-preserved graveyards and clinging precariously to the top of cliffs overlooking the sea, neatly trimmed either by statute and tradition, or a combination of the weather and harsh environmental conditions.

The list of those who make it their business to keep grass in check is long, including ruminant livestock and rabbits, the weather, the local authority, grounds staff, and the gardener in weekend lawn-mowing warrior mode. For a category of plants so ubiquitous – it's rare even for freshly laid turf to include just a single genus, species, or variety of grass, with specific blends of complementary varieties being used to suit different circumstances – it's perhaps unsurprising that we treat grasses with that peculiar kind of disinterest borne of familiarity, only really becoming exercised upon the subject when something seems amiss. It's hard truly to appreciate the unusual beauty of grass when we're societally conditioned to seeing it in an amputated condition, and few of us ever stop to question the uniquely unsustainable nature of the lawn itself. We water and feed it to make the grass grow, then – somewhat perversely – cut most of that new growth off and throw it away. We may compost the cuttings smugly on site but, where space doesn't allow, we must rely upon it being spirited away with the refuse collection. It's a resource-hungry enterprise that can involve water, fertilizer, herbicides, pesticides, and the energy used in their manufacture and transport, to say nothing of that which powers the mower. Ideally, those of us without huge gardens would foreswear our electric and petrol-powered machines, scour boot fairs and garage sales for old push mowers to recondition (the very few models currently in production are lousy) and learn to love a lawn that fades and ripens with the seasons and the sun. But such change requires effort, and to entertain the kind of fascination for a plant that might encourage us to pause for thought is a big ask when we've been taught since childhood continually to knock it back. Perhaps we need to take a step back and spend some quality time in the company of the plants we know as grasses, the better to appreciate their finer qualities.

It might seem hard to believe as we stride across a wide, green sward or kick a ball around with our kids in the park, but that small grass under our boots has a noble pedigree. The botanical family Poaceae to which it belongs numbers among its clan plants that have been essential both to the development and maintenance of human civilization; staple food crops represented by rice and cereals, building materials by bamboos, as well as plants that add to our quality of life in the form of the amenity grasses used in lawns and sports turf, and the ornamental grasses that grace our flowerbeds. To their value to us should be added the ecological worth of each in its natural habitat, though those plants with high economic value tend to be highly hybridized and planted in monocultures in order to simplify husbandry, harvesting, and to maximize yield, all of which reduces the diversity of the wildlife they are able to support.

But pluck a grass plant from the turf and take a good look at it; as so often, given the microscope of our close attention, the ostensibly familiar takes on an air altogether more wrapped about with wonder. In common with other plants, all grasses have roots, stems, and leaves, though the arrangement is not typical, and in time – given the opportunity – flowers and fruit. A fibrous root system produces stems, or culms, about which the long, strap-like leaves wrap themselves, the leaf sheath almost encircling the culm; entirely to the back and sides, arranged in two symmetrically tailored lapels, neatly fastened to the front. A cantilever arrangement that allows the blade of the leaf with its characteristic parallel veins to lean out into space, frequently departing from the stem at an upwardly acute angle until gravity takes a hold and folds the endmost section back towards the ground. Following the best kind of engineering practice, the ligule performs the function of a washer, sitting between sheath and culm to prevent the ingress of water and pathogens and, taking form variously as a membrane or an outgrowth of hairs according to species, can be used as an identifying factor when attempting to differentiate between superficially similar grasses. It seems like a prudent inclusion, and one that reflects on the plant's determination to let nothing frustrate its destiny, for if there's one behaviour that might set grasses aside from other members of the plant kingdom, it's their resilience. The grass's total dedication to growth, often in the face of continued environmental pressure, is due to not only the hard work but the location of its meristems (see *Growth and development*, page 132), the growing points where cell division and multiplication is at its most energetic. The intercalary meristems are found at the base of the leaves and the knuckle-like nodes that punctuate the culm, but the main growing point – the apical meristem – is located at the crown of the plant, close to ground level and below the point at which it might be caught by the scissoring and ripping of ruminant animals' teeth or the blades of the mower. As the seasons progress, the grass moves from a vegetative to a reproductive state, flowering stems are produced and the lowest section of the stem elongates, raising the apical meristem upwards and away from the ground. That this action places the main growing point squarely in the cutting zone can be considered a declaration of intent from a plant that, having flowered, feels itself in need of a rest, and perfectly justified in passing the responsibility for new growth to dormant buds in the crown until spring comes around again.

Staring appreciatively at a single grass is one thing, but having it incorporated en masse into the garden is quite a different prospect. Some of us have been gifted with a lawn – it's customary for a house with a garden to come with such a thing – and whether buying or renting a property that includes within its bounds some access to the ground, there's likely to be a stretch of turf to go with it. The new garden owner may have coveted the feature for some time, or feel little inclination to perpetuate an admittedly green but uninteresting stretch of apparent monoculture, but even the most humble example, unmolested by chemical interference, supports a wealth of life, including a variety of flora adept in creeping along the surface of the ground to avoid the kiss of the mower's blade, or quickly to elevate a lanky flowering stem at the first sign of inattention from the gardener. Hugging the ground there is creeping buttercup and self-heal, speedwell and clovers, ground ivy, bird's-foot trefoil, chickweed, mouse-ear and daisies, while biding their time in low-growing rosettes of foliage are dandelion, hawkbit, ribwort plantain and, glowing emerald bright when the grass is parched and tired, patches of moss, too – all those things that drive a lawn purist to distraction and have them reaching for that combination of fertilizer and poison euphemistically referred to as "weed & feed". But for those of us who thrive on the proximity of diversity and don't want to become a slave to the incessant demands of lawn care, there's the meadow, or at least, the nearest approximation of the thing we can manage in a domestic garden; between the flowerbeds, in a group of troughs on a patio or balcony or even rising from a window box.

Since the definition of a meadow is closely allied to the way the land is managed, simply leaving the lawn a bit long is unlikely quite to be worthy of the name, but this can be our own declaration of intent, a celebration of long grass and flower that defies the strict neatness and order of a traditional style of domestic gardening, whose propensity to make the neighbours tut with its relaxed deportment is offset by the undeniable benefits offered in terms of the richness of the flora and fauna our garden can support. An established meadow can support up to 45 species of grasses, orchids, and other wildflowers per square metre[20], providing cover for ground-nesting lapwings, curlews, skylarks, or yellow wagtails, nectar for butterflies, and pollen for bees, rich populations of invertebrates as well as food for seed-eating birds in late summer and autumn. For most of our gardens, diversity to this degree will remain an ambition, but even the buzzing around a container sown with a simple "meadow mix" – grass seed and wild carrot, field scabious, yarrow, and cornflower – gives a resounding testament to the benefits of incorporating nectar-rich flowers among the grass; while the lawn, left to itself, will begin to augment the community with those plants the land decides to be appropriate in supporting a richer cross-section of life. We can choose to introduce our own selections too, and these are best included as plug plants with their own established root systems, rather than being scattered as seed, which tends to have a higher failure rate. In fact, the more we exert our influence over what should be grown in a meadow, the more frustrating progress can seem to the extent that, for better or worse, new meadow projects are most frequently preceded by a scorched-earth approach to eradicate pernicious weeds, notably thistle and dock, that will compete with the meadow plants for resources. That this approach most often involves the use of precisely the kind of pesticides that have been shown in studies to pose a significant and enduring risk to the health of those bees and pollinators that the meadow is hoping to encourage is an entirely dissonant phenomenon, one to which resolution can only be found in recommending that any new meadow or "wildflower planting" be instituted under a completely organic regime, with all the exasperated dedication that inevitably involves.

As in every part of the garden, the root of our exasperation lies in our own prescriptiveness and an inability to make peace with the notion that the land might simply prefer to produce something other than what we've decided should be grown. It's a difficult balance to strike when we want to introduce long grass and the wonders it can contain into our domestic space. We can't entirely leave nature to it, since a meadow, as already suggested, is a managed habitat, and some intervention is necessary to hold the land at this condition. We can sow yellow rattle, which parasitizes the more vigorous grasses, giving other wildflowers a chance to establish and, if chemicals are to be avoided, cut the whole lot down several times a year, at least initially, in the hope of weakening the undesirables. Out in the wider landscape the meadows will be left to grow until ground-nesting birds have left and late-blooming wildflowers have had a chance to set seed, at which point, the hay cut can take place, severed stems left upon the soil for a day or so in dry weather to allow seeds to fall to the ground and supplement the resident community of perennial plants. Grazing animals can be introduced at this point, their presence significantly cutting down on manual weeding, and their manure restoring any nutrients lost in the harvest.

With our meadow-type plantings on a smaller scale, we're unlikely to find a skylark nesting in a window box or a curlew in the dolly tubs, and few of us have the room to permit a flock of sheep between the hydrangeas, at any time of year. But if a true meadow remains out of reach for most of us, even a nod in its direction brings benefit to the smaller garden and borrowing some of the meadow's principles and attributes accrues a high value. A looser look introduces a blank canvas of long grass across which we are free to draw new lines and routes, delighting in the textural contrast of a mown path through tall wildflowers; a reduction in time and energy spent upon the lawn and, depending upon how fussy we're going to be about quite what grows within, a more relaxed attitude to planting that increases our garden's potential to support a multiplicity of life. A greater appreciation of grasses in their own right – not just the ones we plant in flowerbeds and appreciate for their form, but the kind we find in the lawn, allows both to flourish and flower as we marvel at their texture and their ability to catch the autumn sun and make it sparkle from the thousand tiny lenses held within each plume of feathery flowers. Above all, the song of the meadow forces us to become attuned to the rhythm of the seasons and opens our eyes to the ways in which our interactions with plants impact the wider life of the landscape. Autumn is the time for leaves and long grass, and we are defined as gardeners by what we do with both.

How to PICK AN APPLE

There are so many ways to enjoy an apple. Will you munch it straight from the bowl, pack it away for later in the day, or wait while it bakes, stews, or bubbles away gently in the oven under a crust of crumble or shortcrust pastry in the company of nutmeg, cinnamon, and clove, and the ever-present hope of custard? Apple in hand, still there are matters to decide. How will you attack, hold it, sink your teeth in for the first bite? Will that be daintily or hungrily taken? Nibbled off with decorum, or wrenched away with wolfish delight and accompanying sweet spray of juice? Will there be slurping and sucking, or might you flick open a pocketknife, and cut the fruit into slices? What happens with the core? The pips? The stalk and the blossom? Will you eat the lot, or share the woodier parts with the dog, or the compost?

But first, your apple must be picked.

You can do it quickly. Sometimes you need to. But you should make time to think about what you're doing, at least once in every harvest. You owe this to your apple tree, and to your garden.

A pause. An inhalation…

…hold this a beat or two longer. What follows can all be packed into one eagerly held breath.

An entire season streams into this moment; expectation, promise, sunshine and rain, the buzzing of the bees among spring blossom and the patter of raindrops on fresh green leaves in May, all coalesce and meet in this fat, round fruit, which now you reach for, and cup within your hand.

Take a moment to look at what you hold. It would be unreasonable to expect us to resist eating at least one apple here, at the foot of the tree; one, out of an entire harvest. There's a time and a place for discipline. This is a time for celebration.

Don't tug at the fruit. The tree will give up its bounty when it's good and ready. If there's anything more than the slightest resistance, come back the next day, and try again.

Ever so gently, twist the apple, and listen carefully for the almost imperceptible click as the branch releases its hold on the vessel with its precious cargo; the seeds from which the next generation will spring. Listen carefully still, for…

…almost a sigh. You – or the tree? Sometimes it's difficult to tell.

All season long, this is what it's been about. And now there is a compact between the two of you.

Release the breath and take a bite.

What goes up

The truth is however fast we scrabble to come up with innovative green technologies to repair our world, nature has often got there first. The recyclable solar panel that follows the daily wanderings of the sun through its celestial fields has been around for a while now. There's the type whose photoreceptors are optimized to make the most of available light all year round, but in autumn we find ourselves most concerned with the kind that boasts an impressive level of built-in redundancy; so much so that, in a fully automated process, they are physically jettisoned from the main apparatus at the end of their useful life. With a prescient awareness of the sustainability requirements of modern-day manufacturing, they seem designed to biodegrade where they fall, though the original modelling failed to take into account such variables as paths, patios, lawns, or flowerbeds; items commonly found in the garden though towards which the patent holder maintains a marked antipathy to this day. Should it be considered desirable to move the spent remnants of the season's technology to a different location – a leaf pile by the compost heap perhaps (because we are, of course, talking of leaves), well, that's a job for the gardener. It's as though nature were saying, it's been a perfectly good system for millennia. If you want to stick a lump of concrete in the way, you can deal with the fallout.

We are none of us getting any younger. Doubtless you and I will age with a like degree of effortless grace, striving daily to keep the number of bits that fall off our respective persons to an absolute minimum. It's not the same for a plant, particularly in the case of the deciduously perennial, with whom we'll be concerning ourselves in this chapter owing to the colour and character they lend to the garden in autumn. But while people and animals are flung into the whole ageing thing with neither a dress rehearsal nor certain hope of an encore, these plants get to take a short winter break, roots seemingly steeped in some icy and evasive elixir of life, before starting afresh with youthful vigour come spring. Decrepitude is inevitable for most of us, but a tree or a shrub is better practised at it than a gardener will ever be. Still, we can look on with appreciation and wonder at the annual leaf drop – a process that gifts us, in chronological order, with both spectacle and effort, the one hopefully providing sufficient payment for the other; though there's leaf mould, too, and that lovely soil conditioner should make adequate recompense even for the fussiest among us.

Since budburst in early spring, deciduous leaves have been clinging on for dear life to twig, branch, and stem by the slenderest of appendages, the thin stalk or petiole that provides not only a means of attachment at its base, but a passage through which all energies must flow: xylem vessels bringing nutrient-rich water upwards from the roots, phloem tubes carrying the sugars and starches manufactured in the chloroplasts to wherever they might be needed (see *A free lunch*, page 123). It's two-way traffic for as long as the leaf persists as a functioning unit, but eventually the time comes for even these good and faithful servants to be let go. During the notice period, two related but distinct processes combine to create those conditions in which a plant's leaves find it no longer feasible to resist the persuasive arguments of gravity: senescence and abscission. Neither process is a uniquely foliar affair – petals and fruit also mature, ripen, and age before finding themselves eventually being let go, as do seeds and, to an extent, bark – but both have specific implications for the leaf, its ability to photosynthesize, and the products of that particular chain of reactions.

Ground zero for these critical events is referred to as the abscission zone, the area at which the base of the petiole joins the stem, operated by the plant as a kind of seasonal airlock. But before this closes for good, there is a job of recovery to be done. If a woody plant is to abandon its more tender parts, the better to survive the ravages of winter, it's wise to make one final withdrawal, calling back all investments in terms of nutrients and minerals made over the season, as well as any interest – in the form of carbohydrates – that might have accrued. These are the resources that will fund next year's leafy growth and once stored away over winter in the vaults of root and stem, their erstwhile containers can be let go. Though environmental stress can be a cause of senescence, with plants dropping leaves as a response to drought towards the end of summer, it's the autumnal combination of reduced day length and temperature that sufficiently alters the balance of the plant's growth regulating hormones to usher in this most characteristic of seasonal changes. While biologists continue to explore the details of the interplay between these chemicals – initially abscisic acid was chiefly implicated in the process, though ethylene has now been shown to play a more universally significant role – anyone in possession of a bowl of fruit can watch the effect of a plant hormone upon its contents. This is particularly true of bananas, notable for the generosity with which they release ethylene gas to the atmosphere. There's a point at which a piece of fruit goes from ripe to rotting (a pear, in particular, will spend an age in the bowl in imitation of a green pebble, before sprinting through the distinction while your back is turned), and the presence of ethylene needs to be managed with care if the fruit is not to spoil – the key, at least on a domestic scale, is to keep your bananas separately. This is senescence in action, and something akin to such fruity phenomena occurs within an autumn leaf as its internal architecture degrades, the membranes of cell walls breaking down around the wholesale export of carbohydrates, amino acids, and minerals. It's the dismantling of chlorophyll for parts – not least for that key trace element, magnesium – that reveals the colours of other pigments: the yellow xanthophylls and orange carotenoids present in the leaves all along, while the richest red of anthocyanins begins to develop in tandem with the fading of the green.

By now, the direction of travel through the petiole is distinctly outward bound and, as the tissue of the leaf is denuded of its liquid assets, the same growth-regulating hormone group that plays such a role in the development and direction of roots and shoots during the growing season reveals another of its functions – by making a swift exit. Having spent spring and summer holding back growth in the abscission zone, the onset of senescence sees auxin levels drop and allows for the consequent buildup of waxy suberin, sealing up the base of the petiole and blocking the flow of fluids into and out of the leaf. On the other side of this airlock door, the cells of the petiole continue to deteriorate and, cut off from the rest of the plant, the faintest mechanical encounter or breath of air causes the leaf to release its grip, departing upon the breeze for its final earthbound journey and leaving behind it a small scar at the point of departure – a memory of the leaf that was, and is no more.

There's the unmistakable scent of caramel on the cool afternoon air, perhaps with a hint of vanilla. I emerge from the Moon Gate onto the formal square lawn at the back of the house at Great Comp Garden in Kent, surrounded on all four sides by deep borders stocked with the whole parade of late season daisies – heleniums, dahlias, asters, goldenrod, marigolds, chrysanthemums – interwoven with tall, silvered grasses and underplanted with sedums. But the scent, I know, has nothing to do with these fading blooms, nor with the towering redwood (*Sequoia sempervirens* 'Cantab') that stands sentry beside the steps. It's no subtle hint, no mere suggestion on the breeze, but a full, toffee-on-the-stove onslaught on the olfactory system, and the first time it set my nostrils twitching I was dumbfounded for a while, scouting around the gardens for the source, until I dredged up from some corner of my memory a peculiar feature of the katsura tree, whose leaves contain the chemical compound maltol. Used as an artificial flavour enhancer (with the European food additive number E636) and found naturally in baked bread and barley, when maltol combines with the sugars in the leaf as the chlorophyll breaks down, the tree emits its characteristic smell. And there on the edge of the nursery, presiding over the wooden tables where garden curator William Dyson's salvias proclaimed their flair for bringing colour to the autumnal border, I found the buttery-yellow canopy of the katsura, *Cercidiphyllum japonicum*, whose heart-shaped leaves, when I scooped them off the ground and rubbed them together between my hands, gave off no noticeable scent, though it wafted all about. A small disappointment, since the perfume of the katsura is not universally experienced, I can count myself lucky in being able to detect it at all.

Usually, foliage requires a match to be put to it before the sugary scents are liberated, but the burning of great piles of leaves is a practice thankfully falling out of fashion as being both antisocial and a waste of a wonderful resource for improving the structure of soil (see *Reduce, reuse, recycle*, page 48). Even so, in gardens fortunate enough to count among their number a selection of trees and shrubs – to say nothing of those colourful vines, the Boston ivy and its relative the Virginia creeper, famous for their contribution to the finery of an East Coast fall – there is work here to be done once they have divested themselves of their glory. The epilogue to every autumnal display features a barrage of crinkled confetti, variously blowing about the garden or lying in layers over every horizontal surface and the gardener, wielding leaf rake and barrow, or some noisier combination, must spend a week or two in rearranging the aftermath. Perhaps it wouldn't feel quite so laborious were we to carry out the task within a caramel-scented cloud – oh, for the space to have a katsura tree! Or any tree whose leaves still have the capacity to add particular delight to the experience, even in their fallen state: the scarlet, maple-like hands of the sweet gum, the fiery orange hues of the stag's horn sumach, the traffic light shades of the Persian ironwood, and the gold-brown Vulcan salute of the tulip tree. Even mizzly afternoons under grey October skies may be brightened by the company of such, raked into rows, then rows into piles to be bagged or barrowed to their final resting place by the compost bins, there to undergo over the next few months a final transformation into quantities of deliciously friable leaf mould.

With this as their destiny, even the soggiest of dull brown leaves is worthy of respect and thanks, and the job of collecting them can shift in our contemplation from chore to treasure hunt, with a measure of repetitive aerobic activity thrown in for good measure. For this, I have learned to seek out the lightest leaf rake with the widest head, to master the art of raking on my left side when my right gets tired and that, should arms, knees, or back be unequal to the task, the lawnmower will both collect and mulch the leaves at the same time, providing the lawn isn't too muddy for its wheels to pass across. And that, though blowers are an invaluable aid in shifting leaves from garden furniture and areas of hard landscaping, I prefer to keep my contribution to the choir of power tools within the autumn soundscape to a minimum. Besides which, an experienced gardener with a rake can give a leaf blower a respectable run for its money when it comes to clearing a medium sized lawn of its colourful encumbrances.

And all the while a siren voice murmurs, "Leave them where they fall. Or, if not quite there, blow them onto the beds." It's a tempting thought – after all, we're advocates of returning organic matter to the soil – but little more than an empty promise, a mirage holding out for the hope of rest and refreshment, corners to be cut with no reckoning to be made. But a garden, for the most part, is not woodland, and a thick layer of decomposing leaf litter is inimical to many of the plants that we spend the rest of the year fussing over in the flower beds. Woody perennials may well be able to shrug off a deep and leafy mulch – may even revel in its presence – but the delicate crowns of dainty herbaceous plants are more vulnerable to rot and decay, their fresh new shoots in spring an irresistible draw for the devoted and toothy attention of the slugs that flourish among such delectable decomposition.

Not all gardens experience the fall of autumn leaves in the same way, though those who count their blessings in being spared the annual tidy up might find themselves wondering what else they're missing out upon. Several hours of bending and scraping away with a leaf rake, certainly, and a ready supply of leaf mould that's a solitary single vowel away from "soul conditioner", a description that goes to the heart of the matter. It's these quiet moments spent outside in the fresh air, hands thrust deep into the bodily remains of the garden's recent existence, that bring us viscerally closer to the life of this zone outside our door, this green threshold that gives us the time we need to adjust our face between leaving home and meeting the wider world, this place with its births and deaths and marriages, its rhythms and encores and now, with the leaves before us on the ground, we can choose whether we view our imminent occupation as merely an item of garden administration, or an opportunity for deeper relationship with this living space.

Every garden should have a tree or a shrub, where space permits, and even when it doesn't, quite. Even the smallest of gardens benefit from deciduous autumn colour and, though it makes sense to have that enduring evergreen presence, I want to celebrate the seasons with bright leaves tumbling to the balcony floor, bare stems in winter, and the joy of new buds bursting in spring. I would want dwarf varieties in containers of those plants that bring so much to the garden in all their moods: the Japanese maples, oak-leaved hydrangea, *Rhus typhina* 'Tiger Eyes', all coming at the cost of a little autumnal genuflection, hardly a price beyond our purse. For we can all take a leaf out of the book of our deciduous friends that, irrespective of the season's successes or failures, their own personal degree of growth, and the fruitfulness of their efforts, reach a certain point in the calendar and then, come what may, draw a line under the year and let everything go. For them, their petals, flowers, fruits, and seeds, but mostly, leaves; for us, our hopes, dreams, fears, and concerns – everyone can be subjected to the closest of scrutinies in the hope of finding something that might prove useful in the next attempt, before whatever is left gets consigned to the wind and abandoned to gravity. With wisdom far older than ours, plants have learned the futility of holding on to what no longer serves, and the liberation of letting go, leaving them with the short days and long, dark nights of winter to rest and take stock, before trying again next year. In this, as in so much else, we could all do worse than to think like a plant.

Getting it wrong

Equinox. Night equals day. It sounds like it should be significant for the garden, and it is, arriving in the final full week of September just as the garden is languishing in its fruitful climax. We rise with it on a swell, Nimrod returning home only in place of the spoils of the hunt, we bear dahlias and cosmos and zinnias by the armload, baskets of French beans and trusses of sweet, red tomatoes snipped from the vine, buckets of apples, bowl after bowl of blackberries won from the brambles we endured all year for this harvest, purple-staining hands (and mouths, because who can pick without eating?) and it's now, in this moment toward which everything has been building throughout these past few months, we're brought up short with the realization that this is it, that we have done all we can for one year, and that any effort spent on this ground beyond the harvest will be an investment we make in the next.

A moment, then, to stand and stare upon the crest of a single hill in a rolling landscape, gazing off into the middle distance across the sweep of a wintery valley, the road rising through mist faintly seen on the other side, where you imagine you can just make out the first glimpses of spring. A way off, and much nearer in memory, behind you the road you've travelled; and look how far you've come! And what you've managed to persuade from the ground and – perhaps more importantly – what you haven't quite managed to bring into fruition; those mistakes you've made along the way you might initially be inclined to brush aside if it weren't for your second, and wiser, impulse to hold them in a tight embrace – keeping each individual failure and disappointment captive until it reveals the lesson it has to impart. Because when it comes to gardening, as with any creative enterprise in which you're hoping to grow in confidence and ability, if it's all going right, you're doing something wrong.

That "something", inevitably, is in playing it safe and small.

I might have gotten this wrong, but I think I was gardening before I had a clue what I was doing, when we were still living in a second storey flat in north London. When I bought my lunch from the sandwich shop in Villiers Street and went every day to sit on a bench in the Thames-side gardens on Victoria Embankment – I'm sure I was gardening then, though not a single clod of earth was turned by my hands, nor a solitary leaf disturbed. When I looked intently around and took notice of every development, caressed daily the camouflage bark of the plane trees with a glance, discovered plants I'd not seen before and watched them change as one season blended seamlessly into the next; started with surprise at the sudden appearance of alien, white flower structures on the fatsias in October and passed an approving eye over serried ranks of tulips in spring; when I adopted this narrow strip of ornamental parkland between the bridges of Hungerford and Waterloo as my own for forty minutes five times a week – I think this was gardening, of a sort. It was certainly a very low input method, from my point of view, though I learnt the importance of commitment, and of turning up to do the work – such as it was – whatever the weather. I had an intimate and private connection to this space, my contribution never, much to the relief of regular visitors and tourists alike, the total of the gardening it would receive; there were others who planted and weeded, tidied, and pruned, and turned the place into the kind of garden we've come to expect as a public amenity. But I stubbornly persist in the opinion that this place felt just as gardened by the attention lavished upon it every working day by a sometimes-soggy office worker with his sandwich bag in hand, as by the ministrations of those paid to busy themselves along its paths and among its beds and borders with their leaf blowers, rakes, and barrows filled with brightly coloured bedding plants.

I accept that gardening is more than a spectator sport. But then, while there's so much more to regular, thoughtful observation than looking about you and declaring, "how lovely!" (though this, too, is an undervalued currency, since both gardeners and plants appreciate a little fluffing from time to time), I also have to believe that being a gardener need have nothing to do with the ability to kneel down on a board to weed the edges of a border, or push a barrow, or thrust a spade into the soil; nothing to do with ownership or privilege, access or ability, or else it is open to charges of elitism and ableism, and beyond the reach of the majority. Is it possible I've got this wrong? Surely there's a danger that in so thoroughly democratizing the occupation, the better to ensure universal access, I'm emptying the word of its meaning; if it turns out that we're all gardening simply by virtue of being in a garden, then what's the point of the verb? But no. Have faith. You're not gardening if you're having a picnic or reading the paper, writing a shopping list, or wandering through while chatting to a friend. There is something missing from the equation (you *plus* plants *does not quite equal* garden), but equity is restored when we remember that instructive duo, *intention* and *purpose*, and acknowledge what it would mean to bring them to bear; both upon our attentiveness and the readiness to be fully present in that space, irrespective of the physical control we might actually be able to exert over its contents.

This is all very well, but there's no denying that something in that equation shifts when we needlessly and willingly take on responsibility for a plant. When we introduce a new plant into a garden – a garden that probably didn't need that plant, was almost certainly quite happy growing what it was growing before we arrived with our tastes and our ideas – we make an intervention that raises the stakes, not only for the measure of our prowess as gardeners, but for the plants whose survival now depends upon our vision, skill, and attention. Houseplant enthusiasts, as famously reticent as they can be about taking upon themselves the mantle of "gardener" until they have a patch of ground to tend, often seem to understand this aspect of gardening instinctively. The plants that we nurture in containers are entirely reliant upon the attentions of the gardener, living largely solitary lives, the resources upon which they can draw limited to whatever gets sprayed onto their leaves or poured into their pot. Plants inserted into outdoor communities might have larger reservoirs upon which to draw for sustenance, but even the most well suited to our garden's particular conditions are still relative strangers with few close family ties in the area, each with a relatively short time to put down roots and establish deep connections before the everyday challenges a plant must face can overwhelm its health and energy. It's incumbent upon the gardener, as the one who introduced these plants to this soil, to do everything they can to help them settle in and thrive.

It's a shift concerned not so much with the difference between gardening actively or passively, nor even the matter of responsibility to which I've just alluded. More, it's that defining question of ego, which perhaps more than anything sets us apart from our natural kin; certainly in exerting our own will over our surroundings, but in being prepared to do so in that particularly human way – without a thorough understanding of the complexity of the systems at play, the implications of our action, and the consequences for any dependencies that might be involved. That the adaptability of the natural world allows us to get away with it so often is one of the crowning glories of gardening; that our carelessness and greed can lead us to actions with far reaching and potentially irreversible consequences is its chief tragedy. And so we do what we can, eschewing peat for the destruction that industry wreaks upon a unique habitat, and pesticides for the mayhem they cause in every other; where we must have fertilizer we choose organic but favour seaweed and home-brewed, stinky compost tea, and tread as lightly as we can upon the earth but still there's no escaping it, still we have to make our mark and grow something that wasn't there before. We seize the confidence to suggest some outlandish scheme – a cutting garden! Kitchen garden! A lavender and rosemary parterre! – and see it through to failure or success; to present our idea into the space and risk it all upon a throw of the dice, or at least the weather and the slugs, which amounts to much the same. And this is good – glorious, even – because now we are claiming the agency we have to get it wrong. We are committed, and we will be judged, though most harshly by ourselves.

And this is often where the wheels come off.

We might think it's hard to fail if we never set out to achieve anything. If we keep ourselves small and squash our ambition away into a tiny space within. But if we listen to that voice, not only are we depriving the world of the creativity which only we can bring into being but, in our efforts to avoid failure altogether, we are failing to live life to the full. In the context of the garden, our fearfulness places our ideas and ambitions at peril of never seeing the light of day, for others to experience and enjoy. Our aversion to risk would deny us the knowledge and wonder of discovering what we can achieve in relationship with the world of plants and animals and soil and sky. It's entirely understandable that, notwithstanding the ready availability of advice on every aspect of gardening, it's easy to become overwhelmed by questions of when to prune this or plant that, how to take cuttings, how to sow seeds, when to deadhead, what to do when aphids arrive on your roses. It's easy to become so beset by the fear of doing it wrong that you never quite get going, but really, when you think about it – what's the worst that can happen? In the exceedingly unlikely event that you plant something upside-down – leaves in a hole in the ground, roots waving about in the air – it will quickly expire. But what have you lost? Six quid on a plant, a week or so of wondering why it wasn't looking so hot, and maybe the odd blush. All your future plantings will be exemplary (you wouldn't make that particular mistake twice), and the errors into which you're far more likely to fall will be of greatly less consequence.

Getting it wrong, it transpires, is just about the best instruction we can receive, providing we refuse to become disheartened by the experience and we learn from it what not to do. Temptation tugs on all of us with whispered inducements to minimize our gardening ambitions precisely to avoid the shame of failure, but we could do far worse than to learn from the resilience of the bumblebee who, faced with perpetually moving platforms and unpredictable winds, collides clumsily with each plant she visits once every second, until eventually finding her way to the flower. It might not be elegant, but her persistence pays off; just one lesson the garden offers us for free, and it's right there in the flowerbeds, waiting for us to take note. No bumblebee has ever been put off by getting it "wrong".

Getting it right in the garden – well, that's another thing. Your "right" might not be mine, and mine, I'm fairly sure, will not be yours. But we, and we alone, are the arbiters of our own tastes, and should we choose to garden with a lighter touch than those around us, we have no one to answer to but ourselves. Make no mistake, far from playing it safe, to risk relinquishing control in the garden is a radical act, something that requires a degree of trust in the ability of the natural world to know what it's about that is manifestly absent from the way our lives are run today. To tinker unthinkingly at the edges in an effort to subdue, to faff about with mean, narrow borders, bare earth, and straight lines; to seek the answer to every conundrum the space throws up in the garden centre while keeping the whole depressing lot in check with machines and poisons; to begrudge the moments we spend outside taming the unruliness of our fenced-in wilderness, exporting its richness while importing the depleted reserves of some other patch of the planet – this is the busyness of gardening we have grown accustomed to and, once over the threshold, there's a certain comfort in its inoffensive and unchallenging familiarity. Here is safety in polite conformity, but here too is the destructive smallness from which our gardens are yearning to escape. There is a natural world on our doorstep – a wild, uniquely adaptable, ultimately untamable potency that will continue with, without, or in spite of our contribution, but will surely, I choose to believe, be delighted to welcome us back into meaningful communion; will even entertain the notion of allowing itself to be directed for a while by a sympathetic spirit. All it takes is a degree of humility on our part, and a willingness to pay attention, to stand quietly and learn while the garden does what it does best, and grows.

To the neighbours, our weedy borders might signal how far we are getting it wrong. But we know better. We know that life breeds life, and that the more variety we permit within the space, the more diversity the garden can sustain; that diversity itself introduces resilience, and resilience dares to allow us to hope for the future. We know that when our relationship with the rest of nature is marked on our part by rapt attention and respect, the effort that we put into the garden starts to look less like work, and more like conversation. And that this is how we garden by doing next to nothing.

EPILOGUE

And suddenly it's winter, and the garden has gone to sleep. Misty. Damp. Chilly, rather than bitingly cold – a sign of things to come. We need a good, hard winter…

Oh, but we've been here before. How time flies. A whole year of pausing and breathing; of standing and staring; of stopping to listen and watch, to observe, truly to see; to set aside the need to have all the answers and permit ourselves to be schooled by wonder. A year of thinking like a plant – *roots, shoots, flowers, and fruits* – of letting it go, and letting it grow.

A year of *How to Do the Simple Things* that somehow, some day, we forgot came naturally, or lost the confidence to try and risk getting it wrong, over and over; until we got it right.

And here you are, at the end of a book, and the end (or is it the beginning?) of another chapter for your garden. It's been such fun. An honour. Now you – you, and your wild and wilful partner – you can do this gardening malarkey by yourselves.

But then, of course, you always could.

Index

References

1 King James Bible, Genesis 1:28. (2008). Oxford University Press (Original work published 1769)
2 Horwood, Catherine. (2010). *Gardening Women: Their Stories from 1600 to the Present*. Virago Press.
3 Kazuo Isobe, Hiroaki Oka, Tsunehiro Watanabe, Ryunosuke Tateno, Rieko Urakawa, Chao Liang, Keishi Senoo, Hideaki Shibata. (2018). *High soil microbial activity in the winter season enhances nitrogen cycling in a cool-temperate deciduous forest, Soil Biology and Biochemistry,* Volume 124.
4 Blake, William. (1950). "Auguries of Innocence". *Poets of the English Language*. Eds W.H. Auden and Norman Holmes Pearson. Viking. (Original work published 1863)
5 Shelley, Percy Bysshe. (1818). "Ozymandias". *The Examiner*, no. 524
6 Shelley, Percy Bysshe. (1818). "Ozymandias". *The Examiner*, no. 524
7 Shakespeare, William. (1993). *Romeo and Juliet.* Dover Publications. (Original work published 1597)
8 Jekyll, Gertrude. (1899). *Wood and Garden*, Longmans, Green and Co.
9 Priestley, Joseph. (1774). *Experiments and Observations on Different Kinds of Air.* W. Bowyer and J. Nichols.
10 Mills, Alan; Bonne, Rose; Graboff, Abner. (1961). "I Know An Old Lady". Rand McNally.
11 Wilde, Oscar. (2000). *The Importance of Being Earnest*. Penguin Classics. (Original work published 1898)
12 Plath, Sylvia. (1957). *On the Difficulty of Conjuring Up a Dryad*. Poetry, Vol. 90. Chicago.
13 United States Geological Survey. (2018). https://www.usgs.gov/special-topics/water-science-school/science/evapotranspiration-and-water-cycle Accessed 1 August 2022
14 Bartlett, David. (August 2007). Letter in *The Garden, Journal of the Royal Horticultural Society*
15 Linnaeus, Carl. (1758). *Systema naturae per regna tria naturae : secundum classes, ordines, genera, species, cum characteribus, differentiis, synonymis, locis* (10th ed.). Laurentius Salvius.
16 UK Kent Widlife Trust and Buglife Bugs Matter report 2022 https://www.buglife.org.uk/news/bugs-matter-survey-finds-that-uk-flying-insects-have-declined-by-nearly-60-in-less-than-20-years/ Accessed 1 August 2022
17 Anders Pape Møller. (June 2019). Parallel declines in abundance of insects and insectivorous birds in Denmark over 22 years . Ecology and Evolution. 9 (11): 6581–6587. Accessed 1 August 2022
18 Wallington, Jack. (2019). *Wild About Weeds: Garden Design with Rebel Plants.* Laurence King.
19 Plantlife. (2017). https://www.plantlife.org.uk/uk/about-us/news/real-action-needed-to-save-our-vanishing-meadows Accessed 1 August 2022
20 RSPB. Hay Meadows. https://www.rspb.org.uk/our-work/conservation/conservation-and-sustainability/farming/advice/managing-habitats/hay-meadows/ Accessed 1 August 2022
Page 48 Blake, William. (1950). "Auguries of Innocence". *Poets of the English Language.* Eds W.H. Auden and Norman Holmes Pearson. Viking. (Original work published 1863)
Page 150 King James Bible, Proverbs 22:6 (2008). Oxford University Press (Original work published 1769)
Page 160 Browning, Robert. (1993). "Men and Women". *Everyman*. (Original work published 1855) Davies, W.H. (2015). "Leisure". *Collected Works*. Leopold Classic Library. (Original work published 1911)
Page 240 Keats, John. (1977). "To Autumn". *The Complete Poems*. Penguin Classics. (Original work published 1820)